SpringerBriefs in Statistics

For further volumes:
http://www.springer.com/series/8921

Masanobu Taniguchi · Tomoyuki Amano
Hiroaki Ogata · Hiroyuki Taniai

Statistical Inference for Financial Engineering

Springer

Masanobu Taniguchi
Department of Applied Mathematics
Waseda University
Tokyo
Japan

Tomoyuki Amano
Faculty of Economics
Wakayama University
Wakayama
Japan

Hiroaki Ogata
School of Business Administration,
Faculty of Urban Liberal Arts
Tokyo Metropolitan University
Tokyo
Japan

Hiroyuki Taniai
School of International Liberal Studies
Waseda University
Tokyo
Japan

ISSN 2191-544X ISSN 2191-5458 (electronic)
ISBN 978-3-319-03496-6 ISBN 978-3-319-03497-3 (eBook)
DOI 10.1007/978-3-319-03497-3
Springer Cham Heidelberg New York Dordrecht London

Library of Congress Control Number: 2014934588

Mathematics Subject Classification (2010): 62P05, 91G70

Printed on acid-free paper

Springer is part of Springer Science+Business Media (www.springer.com)

To our families

Preface

This book provides the foundation of statistical inference for financial engineering, which is a huge integration of economics, probability theory, statistics, time series analysis, operation research, etc. Black and Scholes introduced the modern option pricing theory assuming that the price process of an underlying asset follows a geometric Brownian motion. However, empirical studies for the price processes show that they do not follow the geometric Brownian motion, and that they often behave like as non-Gaussian dependent processes. Motivated by this, we investigate what stochastic models can describe the actual financial time series data sufficiently, and how to estimate the proposed models optimally. Concretely, we introduce various stochastic processes, e.g., non-Gaussian linear processes, nonlinear processes, long-memory processes, locally stationary processes, etc. For them we will develop the theory of optimal statistical inference based on local asymptotic normality (LAN), which is due to Le Cam. This book also describes a variety of statistical approaches, e.g., discriminant analysis, empirical likelihood method, control variate method, quantile regression, realized volatility, etc. The philosophy is that financial engineering should be constructed on optimal statistical approaches for plausible stochastic processes.

Chapter 1 discusses time series modeling for financial data, and how to estimate such models optimally. The models include general non-Gaussian vector linear processes, nonlinear time series models, e.g., ARCH, CHARN, etc., and locally stationary processes, etc. Their inference is developed by LAN-based optimal theory. We give an approach based on integral functionals of nonparametric spectral estimators, which are applicable to a lot of problems in financial statistics. Option pricing will be discussed for a class of discretized diffusion processes. Also we address the problem of classification for locally stationary processes, and apply the results to actual financial data by use of dendrogram.

Chapter 2 deals with empirical likelihood approaches for financial data. Since the observations are often supposed to be dependent, we will introduce an estimating function which corresponds to the differential functional of Whittle likelihood, i.e., empirical likelihood in frequency domain. After explaining the asymptotic properties of the frequency domain empirical likelihood estimator, we also introduce the extensions of the empirical likelihood method such as Cressie-Read power-divergence statistic and generalized empirical likelihood. As an application, we consider the generalized empirical likelihood estimation method for the multivariate stable distributions. Several real data analyses are also included.

Chapter 3 gives applications of two statistical methods to the financial time series data. The first one is the control variate method. The control variate method is a method to reduce the variance of estimators by use of some information of another process, which is correlated with the process concerned. This method has been discussed in the case when the data are i.i.d. Since financial time series data are often dependent, we extend this method to dependent data and give the application to financial econometrics. The second one is the instrumental variable method. We introduce a stochastic regression model where the explanatory and disturbance processes are correlated. For this we apply the instrumental variable method to estimate the regression coefficients consistently. We also address the problem of CAPM to our analysis.

Chapter 4 introduces two techniques, which can be utilized in study of financial risks. The first one is the method called Quantile Regression (QR), which can be used to analyze the conditional quantile of financial assets. There, by means of rank-based semiparametrics, we provide the statistically efficient version of QR inference under the autoregressive conditional heteroskedasticity (ARCH). The second technique, the Realized Volatility (RV), estimates the conditional variance, or "volatility" of financial assets. Revealing the fact that its inference can be greatly affected by the existence of additional noise called market microstructure, we introduce and study the asymptotics of some appropriate estimator under the microstructure with ARCH property.

This book is suitable as a professional reference book on finance, statistics, and statistical financial engineering, or a text book for students who specialize these topics.

A part of this book was done in a collaboration between Research Institute for Science and Engineering, Waseda University and Government Pension Investment Fund (GPIF) of Japan. We thank all the members, especially, Prof. Takeru Suzuki and Dr. Takashi Yamashita (GPIF) for their cooperation. Our research was partially supported by the following Japanese Grant-in-Aids: A2324401 (Taniguchi, M.), B22700291 (Ogata, H.), B22700296 (Amano, T.).

Finally, we thank the editors of *SpringerBriefs in Statistics* for their kindness.

January 2013

Masanobu Taniguchi
Tomoyuki Amano
Hiroaki Ogata
Hiroyuki Taniai

Contents

1 Features of Financial Data 1
1.1 Introduction 1
1.2 Time Series Modeling for Financial Data 2
1.3 Optimal Inference for Various Return Processes 19
1.4 Introduction to Time Series Financial Engineering 27
References 37

2 Empirical Likelihood Approaches for Financial Returns 41
2.1 Introduction 41
2.2 Empirical Likelihood Method for i.i.d. Data 42
2.3 Estimation with Frequency Domain Empirical Likelihood for Stationary Processes 43
2.4 Extensions of Empirical Likelihood 46
2.5 GEL for Multivariate Stable Distributions 51
2.6 Appendix 55
2.6.1 Proof of Theorem 2.2 57
2.6.2 Proof of Theorem 2.3 60
References 63

3 Various Methods for Financial Engineering 65
3.1 Introduction 65
3.2 Control Variate Methods for Financial Data 66
3.3 Statistical Estimation for Stochastic Regression Models with Long Memory Dependence 74
References 83

4 Some Techniques for ARCH Financial Time Series 85
4.1 Introduction 85
4.2 Quantile Regression and Its Semiparametric Efficiency for ARCH Series 86
4.2.1 Model, Estimators, and Some Asymptotics 86
4.2.2 Semiparametrically Efficient Inference 89
4.2.3 Numerical Studies 95

4.3 Asymptotics of Realized Volatility with Non-Gaussian ARCH(∞) Microstructure Noise . . . 102
4.3.1 Model, Estimators, and Main Results . . . 103
4.3.2 Numerical Studies . . . 109
4.4 Appendix . . . 112
4.4.1 Proof of Theorem 4.1 . . . 112
References . . . 115

Index . . . 117

Chapter 1
Features of Financial Data

Abstract This chapter discusses actual features of financial time series data, and how to model them statistically. Because the mechanism of financial market is obviously complicated, modeling for financial time series is difficult. For this, first, we look at some empirical characteristics of financial data. Then, we review and examine various time series models (e.g., ARCH, general linear process, non-stationary process, etc.), which show plausibility. Their estimation theory is provided in a unified fashion. Optimality of the estimation and testing, etc., is described based on the local asymptotic normality (LAN) due to Le Cam. The theory and models are very general and modern.

Keywords Nonlinear time series models · Vector linear processes · LAN approach · Option pricing · Classification for time series · Locally stationary processes

1.1 Introduction

In this chapter we address the problems of modeling for financial time series data and the optimal inference for various models. Then an introduction to financial engineering based on time series analysis is provided.

Section 1.2 studies statistical features of financial returns. They naturally lead to nonlinear time series models, e.g., ARCH, GARCH, EGARCH, CHARN, etc. If we assume their stationarity, Wold decomposition theorem shows that they can be expressed as linear processes. For general non-Gaussian vector-valued linear processes, we develop the asymptotic theory for Whittle estimators and integral functional of nonparametric spectral density estimators. Then the problems of h-step ahead prediction and testing a strength of causality are discussed.

Lucien LeCam established one of the most important foundations of the general statistical asymptotic theory (e.g., see LeCam (1986)). He introduced the concept of local asymptotic normality (LAN) for the likelihood ratio of general statistical model

M. Taniguchi et al., *Statistical Inference for Financial Engineering*,
SpringerBriefs in Statistics, DOI: 10.1007/978-3-319-03497-3_1,

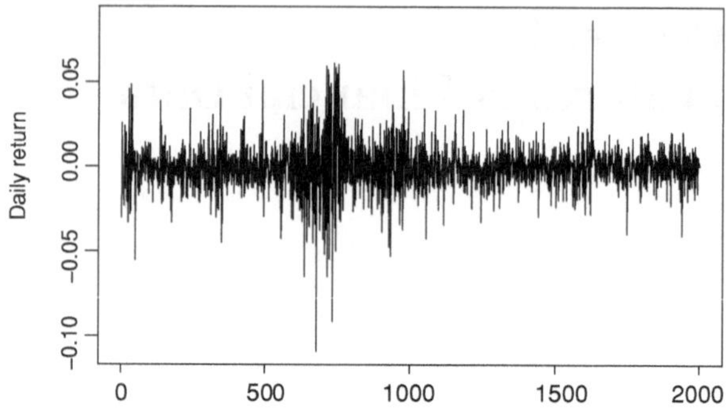

Fig. 1.1 Daily log return of IBM $\{X_t\}$

in i.i.d. case. Beautiful thing is that once LAN is proved, the asymptotic optimality of estimators and tests is described in terms of the LAN property.

Section 1.3 introduces an important class of nonlinear time series models, called ARCH(∞) with stochastic mean (ARCH(∞)-SM) models, which includes the usual ARCH and GARCH as special cases. For this class, we establish the LAN, and describe the framework of the optimal inference based on LAN, proving the optimality of MLE. This stream of discussion is extended to a very general class of conditional heteroscedastic autoregressive nonlinear (CHARN) models and nonstationary processes, called locally stationary processes.

Section 1.4 develops the arguments of financial engineering based on time series analysis. Fundamental concepts, i.e., arbitrage-free, self-financing portfolio, completeness, etc., are explained. We discuss the pricing a European call option for discretized diffusion returns. Classification method of locally stationary processes is introduced. Using a distance between their nonparametric spectral estimators, we executed the hierarchical clustering for daily log-stock returns of 13 companies. The dendrogram classifies the type of industry clearly, which implies that the method will be useful for the problem of credit rating.

Because this book is mainly based on the research papers of the authors, it is recommendable to refer Gouriéroux and Jasiak (2001) for financial econometrics, Van der Vaart (1998) for the LAN approach and Brockwell and Davis (1991) for general time series analysis.

1.2 Time Series Modeling for Financial Data

We begin by looking at actual stock price data. Figure 1.1 plots the daily log return of IBM from October 16, 2003 to September 26, 2011. Write the observed stretch as $X_1, X_2, \ldots, X_n$. As a fundamental analysis we often examine the behavior of the sample autocorrelation function (SACF):

$$\hat{\rho}_{X_t}(l) \equiv \frac{\sum_{t=1}^{n-l}(X_{t+l} - \overline{X}_n)(X_t - \overline{X}_n)}{\sum_{t=1}^{n}(X_t - \overline{X}_n)^2} \tag{1.1}$$

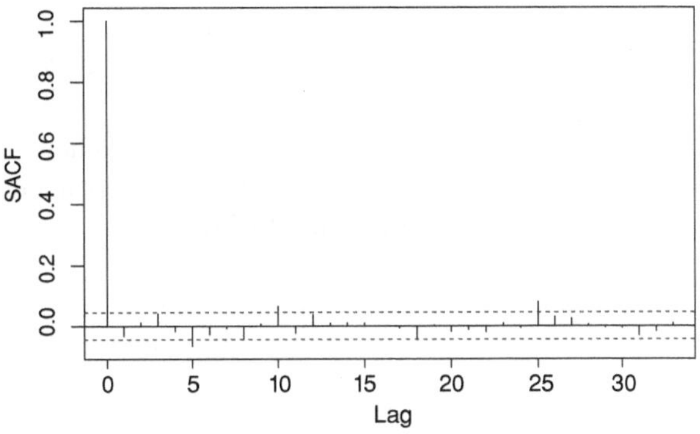

Fig. 1.2 SACF of $\{X_t\}$

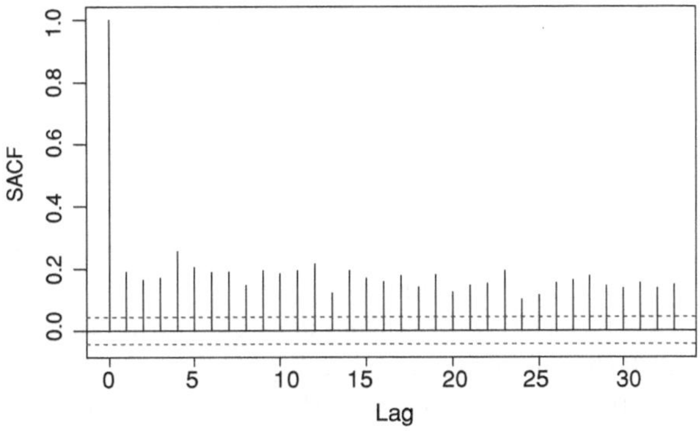

Fig. 1.3 SACF of $\{X_t^2\}$

where $\bar{X}_n \equiv n^{-1}\sum_{t=1}^{n} X_t$. The SACF $\hat{\rho}_{X_t}(l)$ for $\{X_t\}$ is shown in Fig. 1.2. The SACF $\hat{\rho}_{X_t^2}(l)$ for square transformed data X_t^2 is shown in Fig. 1.3. From (1.1), $\hat{\rho}_{X_t}(l)$ shows a strength of interrelation between X_{t+l} and X_t, hence, if X_t's are mutually independent or uncorrelated, $\hat{\rho}_{X_t}(l)$, $(l \neq 0)$, will be near zero. Figures 1.2 and 1.3, respectively, suggest that X_t's are almost uncorrelated, and that X_t^2's are not uncorrelated, which leads to the following conclusion:

X_t's are not mutually independent and the distribution of $\{X_t\}$ is non-Gaussian.

Observing these symptoms of financial data, Engle (1982) proposed an autoregressive conditional heteroscedastic model (ARCH(q)), which is defined as

$$\begin{cases} E(X_t|\mathcal{F}_{t-1}) = 0 \quad a.e., \\ Var(X_t|\mathcal{F}_{t-1}) = a_0 + \sum_{j=1}^{q} a_j X_{t-j}^2 \quad a.e., \end{cases} \tag{1.2}$$

where $\mathcal{F}_{t-1}$ is σ-algebra generated by $\{X_{t-1}, X_{t-2}, \ldots\}$ and $a_0 > 0$, $a_j \geq 0$, $j = 1, \ldots, q$. A concrete representation of it is given by

$$\begin{cases} X_t = \epsilon_t \sqrt{h_t}, \\ h_t = a_0 + \sum_{j=1}^{q} a_j X_{t-j}^2, \end{cases} \tag{1.3}$$

where $\{\epsilon_t\}$ is a sequence of i.i.d.$(0, \sigma^2)$ random variables. Evidently, X_t's are uncorrelated, and it is seen that they are dependent and non-Gaussian generally. Bollerslev (1986) generalized ARCH(q) to

$$\begin{cases} E(X_t|\mathcal{F}_{t-1}) = 0 \quad a.e., \\ Var(X_t|\mathcal{F}_{t-1}) \equiv a_0 + \sum_{j=1}^{q} a_j X_{t-j}^2 + \sum_{j=1}^{p} b_j h_{t-j}^2 \quad a.e., \end{cases} \tag{1.4}$$

where $a_0 > 0$, $a_j \geq 0$, $j = 1, \ldots, q$, $b_j \geq 0$, $j = 1, \ldots, p$, which is called a generalized autoregressive conditional heteroscedastic model (GARCH(p, q)). Further, generalizing ARCH and GARCH, Giraitis et al. (2000) introduced the ARCH(∞) defined by

$$\begin{cases} X_t = \epsilon_t \sqrt{h_t} \\ h_t = a_0 + \sum_{j=1}^{\infty} a_j X_{t-j}^2 \end{cases} \tag{1.5}$$

where $a_0 > 0$, $a_j \geq 0$, $j = 1, \ldots$, $\{\epsilon_t\}$ is a sequence of i.i.d. random variables, and ϵ_t is $\mathcal{F}_t$-measurable and independent of $\mathcal{F}_{t-1}$.

Empirically, it is known that stock returns are negatively correlated with changes in returns volatility, i.e., volatility tends to rise in response to "bad news" and to fall in response to "good news." ARCH and GARCH models cannot describe this aspect. To allow the asymmetric effect, Nelson (1991) proposed the exponential GARCH model(EGARCH(p, q)):

$$\begin{cases} X_t = \epsilon_t \cdot \sigma_t \\ \log \sigma_t^2 = a_0 + \sum_{j=1}^{p} a_j \frac{|X_{t-j}| + \gamma_j X_{t-j}}{\sigma_{t-j}} + \sum_{j=1}^{q} b_j \log \sigma_{t-j}^2 \end{cases} \tag{1.6}$$

where a_j, b_j, γ_j are unknown parameters, and are allowed to be negative unlike ARCH and GARCH parameters. If $a_j \gamma_j < 0$, $j = 1, \ldots, p$, then we can understand the asymmetry of EGARCH.

Recently, the following stochastic volatility model(SV(m)) has been introduced, and is defined by

$$\begin{cases} X_t = \sigma_t \epsilon_t \\ \log \sigma_t^2 - \alpha_1 \log \sigma_{t-1}^2 - \cdots - \alpha_m \log \sigma_{t-m}^2 = \alpha_0 + v_t \end{cases} \tag{1.7}$$

where $\{\epsilon_t\} \sim$ i.i.d.$(0, 1)$, $\{v_t\} \sim$ i.i.d.$(0, \sigma_v^2)$, and $\{\epsilon_t\}$ and $\{v_t\}$ are mutually independent.

Incorporating ARCH with SV, Härdle et al. (1998) introduced the following conditional heteroscedastic autoregressive nonlinear model (denoted by CHARN):

$$\boldsymbol{X}_t = \boldsymbol{F}_\theta(\boldsymbol{X}_{t-1}, \ldots, \boldsymbol{X}_{t-p}) + \boldsymbol{H}_\theta(\boldsymbol{X}_{t-1}, \ldots, \boldsymbol{X}_{t-q})\boldsymbol{\epsilon}_t \tag{1.8}$$

where $\boldsymbol{F}_{\boldsymbol{\theta}} : \mathbb{R}^{mp} \to \mathbb{R}^m$ and $\boldsymbol{H}_\theta : \mathbb{R}^{mq} \to \mathbb{R}^m \times \mathbb{R}^m$, are measurable functions depending on unknown parameter $\boldsymbol{\theta} \in \boldsymbol{\Theta} \subset \mathbb{R}^r$, $\{\boldsymbol{X}_t\}$ and $\{\boldsymbol{\epsilon}_t\}$ are sequences of m-dimensional random vectors, $\{\boldsymbol{\epsilon}_t\} \sim$ i.i.d.$(\boldsymbol{0}, \boldsymbol{V})$ and $\boldsymbol{\epsilon}_t$ is independent of $\{\boldsymbol{X}_s, s < t\}$. Although CHARN has been used for financial data, Kato et al. (2006) applied this model to analysis of brain and muscular waves. Hence, CHARN models are very general.

Besides the above, a lot of nonlinear time series models have been proposed. Among them, ARCH(∞) and CHARN models are typical and general. If we discuss estimation theory for nonlinear time series, their stationarity is one of the most fundamental assumptions. Lu and Jiang (2001) gave a sufficient condition for strict stationarity of CHARN model (1.8). Giraitis et al. (2000) gave sufficient conditions for strict stationarity and second-order stationarity of ARCH(∞) model (1.5).

For second-order stationary processes, we have the following convenient result (for proof, e.g., Hannan (1970), p.137).

Proposition 1.1 (Wold decomposition Theorem) *Any zero-mean second-order stationary process $\{X_t\}$ can be represented in the form of linear process:*

$$X_t = \sum_{j=0}^{\infty} a_j u_{t-j} + \eta_t, \tag{1.9}$$

where $a_0 = 1$, $\sum_{j=0}^{\infty} a_j^2 < \infty$, and

(i) $E\{u_s u_t\} = \delta(t-s)\sigma^2$
(ii) $E\{u_s \eta_t\} = 0$, *for any $s, t \in \mathbb{Z}$*
(iii) *η_t is purely deterministic, i.e., $\eta_t = E\{\eta_t | X_{t-1}, X_{t-2}, \ldots\}$.*

In view of Proposition 1.1, the class of linear processes is a sufficiently rich one, which contains classes of various stationary nonlinear processes. Thus, in what follows, we provide some fundamentals of linear processes.

Let $\{\boldsymbol{X}(t) = (X_1(t), \ldots, X_m(t))'; t \in \mathbb{Z}\}$ be generated by

$$\boldsymbol{X}(t) = \sum_{j=0}^{\infty} \boldsymbol{A}(j)\boldsymbol{U}(t-j), \quad t \in \mathbb{Z} \tag{1.10}$$

where $\{\boldsymbol{U}(t) = (u_1(t), \ldots, u_m(t))'\}$ is a sequence of m-vector random variables satisfying

$$E\{\boldsymbol{U}(t)\} = \boldsymbol{0}, \quad E\{\boldsymbol{U}(t)\boldsymbol{U}(s)'\} = \boldsymbol{0} \ \ (t \neq s)$$

and

$$\mathrm{Var}\{\boldsymbol{U}(t)\} = \boldsymbol{K} = \{K_{ab} : a, b = 1, \dots, m\} \quad \text{for all } t \in \mathbb{Z}.$$

Here, $\boldsymbol{A}(j) = \{A_{ab}(j); a, b = 1, \dots, m\}$, $j \in \mathbb{Z}$, are $m \times m$ matrices with $\boldsymbol{A}(0) = \boldsymbol{I}_m$. We set

Assumption 1.1 $\sum_{j=0}^{\infty} \mathrm{tr}\{\boldsymbol{A}(j)\boldsymbol{K}\boldsymbol{A}(j)'\} < \infty$.

Then, $\{\boldsymbol{X}(t)\}$ becomes a second-order stationary process with spectral density matrix

$$\boldsymbol{f}(\lambda) = \{f_{ab}(\lambda); a, b = 1, \dots, m\} = (2\pi)^{-1}\boldsymbol{A}(\lambda)\boldsymbol{K}\boldsymbol{A}(\lambda)^*,$$

where $\boldsymbol{A}(\lambda) = \sum_{j=0}^{\infty} \boldsymbol{A}(j)e^{ij\lambda}$.

In the inference for (1.10), we often deal with the integral functional of periodogram matrix $\boldsymbol{I}_n(\lambda)$:

$$F_n \equiv \int_{-\pi}^{\pi} \mathrm{tr}\{\boldsymbol{\phi}(\lambda)\boldsymbol{I}_n(\lambda)\}d\lambda, \tag{1.11}$$

where $\boldsymbol{\phi}(\lambda)$ is an $m \times m$ matrix-valued continuous function on $[-\pi, \pi]$, and

$$\boldsymbol{I}_n(\lambda) = (2\pi n)^{-1}\left\{\sum_{t=1}^{n} \boldsymbol{X}(t)e^{it\lambda}\right\}\left\{\sum_{t=1}^{n} \boldsymbol{X}(t)e^{it\lambda}\right\}^*.$$

To describe the asymptotics of $\sqrt{n}(F_n - F)$, where $F = \int_{-\pi}^{\pi} \mathrm{tr}\{\boldsymbol{\phi}(\lambda)\boldsymbol{f}(\lambda)\}d\lambda$, we mention two approaches. The first approach is due to Hosoya and Taniguchi (1982). Consider the linear process (1.10). Denote by $\mathcal{F}_t$ the σ-algebra generated by $\{\boldsymbol{U}(s); s \le t\}$.

Assumption *(HT)*

(i) $\{\boldsymbol{U}(t)\}$ is fourth-order stationary, and the fourth-order cumulants $c^{U}_{a_1 \dots a_4}(t_1, t_2, t_3) \equiv \mathrm{cum}\{u_{a_1}(0), u_{a_2}(t_1), u_{a_3}(t_2), u_{a_4}(t_3)\}$, $(a_1, \dots, a_4 = 1, \dots, m)$, satisfy

$$\sum_{t_1, t_2, t_3 = -\infty}^{\infty} \left| c^{U}_{a_1 \dots a_4}(t_1, t_2, t_3) \right| < \infty.$$

(ii) For each a_1, a_2 and s,

$$\mathrm{Var}[E\{u_{a_1}(t)u_{a_2}(t+s)|\mathcal{F}_{t-\tau}\} - \delta(s)K_{a_1a_2}] = O(\tau^{-2-\epsilon}) \quad \text{for } \epsilon > 0,$$

uniformly in t.

(iii) For each a_1, a_2, a_3 and a_4,

$$\begin{aligned} &E|E\{u_{a_1}(t_1)u_{a_2}(t_2)u_{a_3}(t_3)u_{a_4}(t_4)|\mathcal{F}_{t_1-\tau}\} - E\{u_{a_1}(t_1)u_{a_2}(t_2)u_{a_3}(t_3)u_{a_4}(t_4)\}| \\ &\quad = O(\tau^{-1-\eta}), \end{aligned}$$

uniformly in t_1, where $t_1 \le t_2 \le t_3 \le t_4$ and $\eta > 0$.

(iv) For any $\rho > 0$ and for any integer $L \ge 0$, there exists $B_\rho > 0$ such that

$$E[T(n,s)^2 I\{T(n,s) > B_\rho\}] < \rho,$$

uniformly in n and s, where

$$T(n,s) = \left[\sum_{a_1,a_2=1}^{n}\sum_{r=0}^{L}\left\{\sum_{t=1}^{n}\frac{u_{a_1}(t+s)u_{a_2}(t+s+r) - \delta(r)K_{a_1a_2}}{\sqrt{n}}\right\}^2\right]^{\frac{1}{2}},$$

and $I\{\cdot\}$ is the indicator function of $\{\cdot\}$.

(v) $f(\lambda) \in Lip(\alpha)$, the Lipschitz class of degree α, $\alpha > 1/2$.

In the literature, a higher order martingale difference condition for $\{\boldsymbol{U}(t)\}$ is imposed, i.e.,

$$\begin{aligned} E\{u_{a_1}(t)|\mathcal{F}_{t-1}\} = 0 \quad a.s., \quad & E\{u_{a_1}(t)u_{a_2}(t)|\mathcal{F}_{t-1}\} = K_{a_1a_2} \quad a.s., \\ & E\{u_{a_1}(t)u_{a_2}(t)u_{a_3}(t)|\mathcal{F}_{t-1}\} = \gamma_{a_1a_2a_3} \quad a.s., \quad \text{etc.} \end{aligned} \tag{1.12}$$

(e.g., Dunsmuir and Hannan (1976)). The assumption (HT) (ii) and (iii) mean a kind of "asymptotically" higher order martingale difference condition, which is more natural than (1.12).

To elucidate the asymptotics of $\sqrt{n}(F_n - F)$, Hosoya and Taniguchi (1982) showed:

$\sqrt{n}(F_n - F)$ is approximated by a finite linear combination of

$$\sqrt{n}\left\{\frac{1}{n}\sum_{t=1}^{n} X_{a_1}(t)X_{a_2}(t+s) - E\{X_{a_1}(t)X_{a_2}(t+s)\}\right\} \tag{1.13}$$

where $s = 0, 1, \ldots, n-1, a_1, a_2 = 1, \ldots, m$.

Recalling that $\{\boldsymbol{X}(t)\}$ is generated by (1.10), we can see that (1.13) is approximated by a finite linear combination of

$$\frac{1}{\sqrt{n}}\left\{\sum_{t=1}^{n} u_{a_1}(t)u_{a_2}(t+l) - \delta(l)K_{a_1a_2}\right\}, \quad l \le s. \tag{1.14}$$

Applying Brown's central limit theorem (Brown (1971)) to (1.14), Hosoya and Taniguchi (1982) elucidated the asymptotics of $\sqrt{n}(F_n - F)$.

The second approach is due to Brillinger (2001). Assume that $\{\boldsymbol{X}(t)\}$ is strictly stationary and all the moments exist.

Let

$$c^{\boldsymbol{X}}_{a_1\ldots a_k}(t_1,\ldots,t_{k-1}) = \mathrm{cum}\{X_{a_1}(0), X_{a_2}(t_1),\ldots,X_{a_k}(t_{k-1})\},$$

for $a_1,\ldots,a_k = 1,\ldots,m$; $k = 2,3,\ldots$.

Assumption (*B*) For each $j = 1,2,\ldots,k-1$ and any k-tuple $a_1,a_2,\ldots,a_k$ we have

$$\sum_{t_1,\ldots,t_{k-1}=-\infty}^{\infty} (1+|t_j|)|c^{\boldsymbol{X}}_{a_1\ldots a_k}(t_1,\ldots,t_{k-1})| < \infty, \quad k = 2,3,\ldots. \tag{1.15}$$

Under Assumption (B), Brillinger (2001) evaluated the Jth order cumulant $c_n(J)$ of $\sqrt{n}(F_n - F)$, and showed that $c_n(J) \to 0$ as $n \to \infty$ for all $J \ge 3$, which implies the asymptotic normality of $\sqrt{n}(F_n - F)$.

Summarizing the two approaches, we have

Proposition 1.2 (Hosoya and Taniguchi (1982), Brillinger (2001)) *Suppose that one of the following conditions (HT) and (B) holds:*
(HT) $\{\boldsymbol{X}(t) : t \in \mathbb{Z}$ *is generated by* (1.10) *and satisfies Assumption (HT).*
(B) $\{\boldsymbol{X}(t) : t \in \mathbb{Z}$ *is strictly stationary, and satisfies Assumption (B).*
Let $\boldsymbol{\phi}_j(\lambda)$, $j = 1,\ldots,q$, *be* $m \times m$ *matrix-valued continuous functions on* $[-\pi,\pi]$ *such that* $\boldsymbol{\phi}_j(\lambda) = \boldsymbol{\phi}_j(\lambda)^*$ *and* $\boldsymbol{\phi}_j(-\lambda) = \boldsymbol{\phi}_j(\lambda)'$. *Then*

(i) for each $j = 1,\ldots,q$,

$$\int_{-\pi}^{\pi} \mathrm{tr}\{\boldsymbol{\phi}_j(\lambda)\boldsymbol{I}_n(\lambda)\}d\lambda \xrightarrow{p} \int_{-\pi}^{\pi} \mathrm{tr}\{\boldsymbol{\phi}_j(\lambda)\boldsymbol{f}(\lambda)\}d\lambda \tag{1.16}$$

as $n \to \infty$,

(ii) the quantities

$$\sqrt{n}\int_{-\pi}^{\pi} \mathrm{tr}[\boldsymbol{\phi}_j(\lambda)\{\boldsymbol{I}_n(\lambda) - \boldsymbol{f}(\lambda)\}]d\lambda, \quad j = 1,\ldots,q$$

have, asymptotically, a normal distribution with zero-mean vector and covariance matrix $\boldsymbol{V}$ *whose (j,l)th element is*

$$4\pi \int_{-\pi}^{\pi} \mathrm{tr}\{\boldsymbol{f}(\lambda)\boldsymbol{\phi}_j(\lambda)\boldsymbol{f}(\lambda)\boldsymbol{\phi}_l(\lambda)\}d\lambda$$

$$+ 2\pi \sum_{r,t,u,v=1}^{m} \int\int_{-\pi}^{\pi} \phi_{rt}^{(j)}(\lambda_1)\phi_{uv}^{(l)}(\lambda_2) Q_{rtuv}^{X}(-\lambda_1, \lambda_2, -\lambda_2)\, d\lambda_1 d\lambda_2.$$

where $\phi_{rt}^{(j)}(\lambda)$ *is the (r,t)th element of* $\boldsymbol{\phi}_j(\lambda)$*, and*

$$Q_{rtuv}^{X}(\lambda_1, \lambda_2, \lambda_3) = (2\pi)^{-3} \sum_{t_1,t_2,t_3=-\infty}^{\infty} \exp\{-i(\lambda_1 t_1 + \lambda_2 t_2 + \lambda_3 t_3)\} c_{rtuv}^{X}(t_1, t_2, t_3).$$

Let us think about the case when $m = 1$ (scalar process), and $\{\boldsymbol{X}(t)\}$ is a Gaussian process. In the case we write $\boldsymbol{X}(t)$, $\boldsymbol{I}_n(\lambda)$, $\boldsymbol{f}(\lambda)$ by $X(t)$, $I_n(\lambda)$ and $f(\lambda)$, respectively. Let $\boldsymbol{X} = (X(1), X(2), \ldots, X(n))'$ be a partial realization of $\{X(t) : t \in \mathbb{Z}\}$. Denote the probability density function of $\boldsymbol{X}$ by $p(\boldsymbol{X})$. Under appropriate regularity conditions, Liggett (1971) showed

$$\frac{1}{n} \log\{p(\boldsymbol{X})\} = -\log 2\pi - \frac{1}{4\pi} \int_{-\pi}^{\pi} \left[\log f(\lambda) + \frac{I_n(\lambda)}{f(\lambda)} \right] d\lambda + o_p(1). \tag{1.17}$$

Hence, if the spectral density is parameterized by q-dimensional unknown parameter $\boldsymbol{\theta} \in \boldsymbol{\Theta} \subset \mathbb{R}^q$, an approximated maximum likelihood estimator $\hat{\boldsymbol{\theta}}$ may be defined by

$$\hat{\boldsymbol{\theta}} \equiv \arg\min_{\boldsymbol{\theta} \in \boldsymbol{\Theta}} \int_{-\pi}^{\pi} \{\log f_{\boldsymbol{\theta}}(\lambda) + f_{\boldsymbol{\theta}}(\lambda)^{-1} I_n(\lambda)\} d\lambda. \tag{1.18}$$

Returning to the original setting (1.10), we may understand that the multivariate version of the right-hand side integral part in (1.18) is

$$D(\boldsymbol{f}_{\boldsymbol{\theta}}, \boldsymbol{I}_n) \equiv \int_{-\pi}^{\pi} [\log\det \boldsymbol{f}_{\boldsymbol{\theta}}(\lambda) + \mathrm{tr}\{\boldsymbol{f}_{\boldsymbol{\theta}}(\lambda)^{-1} \boldsymbol{I}_n(\lambda)\}] d\lambda. \tag{1.19}$$

Let

$$\hat{\boldsymbol{\theta}}_{QGML} \equiv \arg\min_{\boldsymbol{\theta} \in \boldsymbol{\Theta}} D(\boldsymbol{f}_{\boldsymbol{\theta}}, \boldsymbol{I}_n), \tag{1.20}$$

which is called a quasi-Gaussian maximum likelihood estimator. Although in the derivation of (1.17), we used Gaussianity of the process, the assumption of Gaussianity may be dropped when we use $\hat{\boldsymbol{\theta}}_{QGML}$.

Let $\mathcal{F} = \{\boldsymbol{f}_{\boldsymbol{\theta}}(\lambda) : \boldsymbol{\theta} \in \boldsymbol{\Theta} \subset \mathbb{R}^q\}$ be a family of parametric spectral density models, which may not contain the true model $\boldsymbol{f}(\lambda)$. We fit $\boldsymbol{f}_{\boldsymbol{\theta}}(\lambda) \in \mathcal{F}$ to $\boldsymbol{f}(\lambda)$ by the criterion $D(\boldsymbol{f}_{\boldsymbol{\theta}}, \boldsymbol{f})$ defined by (1.19) replacing $\boldsymbol{I}_n$ by $\boldsymbol{f}$. This setting is very natural for practical situations, and has expectedly wide applications. Let

$$\underline{\boldsymbol{\theta}} = \arg\min_{\boldsymbol{\theta} \in \boldsymbol{\Theta}} D(\boldsymbol{f}_{\boldsymbol{\theta}}, \boldsymbol{f}), \tag{1.21}$$

which is called a quasi-true value of the parameter. If $\boldsymbol{f}_{\boldsymbol{\theta}}(\lambda)$ is sufficiently smooth with respect to $\boldsymbol{\theta}$, and if $\hat{\boldsymbol{\theta}}_{QGML}$ satisfies

$$\frac{\partial}{\partial \boldsymbol{\theta}} D(\boldsymbol{f}_{\boldsymbol{\theta}}, \boldsymbol{I}_n)|_{\boldsymbol{\theta} = \hat{\boldsymbol{\theta}}_{QGML}} = \mathbf{0}, \tag{1.22}$$

then Taylor's expansion of (1.22) at $\underline{\boldsymbol{\theta}}$ leads to

$$\mathbf{0} = \frac{\partial}{\partial \boldsymbol{\theta}} D(\boldsymbol{f}_{\underline{\boldsymbol{\theta}}}, \boldsymbol{I}_n) + \frac{\partial^2}{\partial \boldsymbol{\theta} \partial \boldsymbol{\theta}'} D(\boldsymbol{f}_{\underline{\boldsymbol{\theta}}}, \boldsymbol{I}_n)(\hat{\boldsymbol{\theta}}_{QGML} - \underline{\boldsymbol{\theta}}) + \cdots. \tag{1.23}$$

If (1.21) implies

$$\frac{\partial}{\partial \boldsymbol{\theta}} D(\boldsymbol{f}_{\boldsymbol{\theta}}, \boldsymbol{f})|_{\boldsymbol{\theta} = \underline{\boldsymbol{\theta}}} = \mathbf{0}, \tag{1.24}$$

then, it is seen from (1.23) that

$$\sqrt{n}(\hat{\boldsymbol{\theta}}_{QGML} - \underline{\boldsymbol{\theta}}) \approx -\left[\frac{\partial^2}{\partial \boldsymbol{\theta} \partial \boldsymbol{\theta}'} D(\boldsymbol{f}_{\underline{\boldsymbol{\theta}}}, \boldsymbol{I}_n)\right]^{-1} \sqrt{n}\left[\frac{\partial}{\partial \boldsymbol{\theta}} D(\boldsymbol{f}_{\underline{\boldsymbol{\theta}}}, \boldsymbol{I}_n) - \frac{\partial}{\partial \boldsymbol{\theta}} D(\boldsymbol{f}_{\underline{\boldsymbol{\theta}}}, \boldsymbol{f})\right]. \tag{1.25}$$

Because the derivatives of $D(\boldsymbol{f}_{\underline{\boldsymbol{\theta}}}, \boldsymbol{I}_n)$ are integral functionals of $\boldsymbol{I}_n$, hence, application of Proposition 1.2 to (1.25) yields,

Proposition 1.3 (Hosoya and Taniguchi (1982)) *Suppose that* $\underline{\boldsymbol{\theta}}$ *exists uniquely and lies in* Int$\boldsymbol{\Theta}$, *and that*

$$\boldsymbol{M}_f \equiv \int_{-\pi}^{\pi} \left[\frac{\partial^2}{\partial \boldsymbol{\theta} \partial \boldsymbol{\theta}'} \mathrm{tr}\left\{\boldsymbol{f}_{\boldsymbol{\theta}}(\lambda)^{-1} \boldsymbol{f}(\lambda)\right\} + \frac{\partial^2}{\partial \boldsymbol{\theta} \partial \boldsymbol{\theta}'} \log\det \boldsymbol{f}_{\boldsymbol{\theta}}(\lambda)\right]_{\boldsymbol{\theta} = \underline{\boldsymbol{\theta}}} d\lambda$$

is a nonsingular matrix. Then, under Assumption(HT), it holds that, as $n \to \infty$,

(i) $\hat{\boldsymbol{\theta}}_{QGML} \xrightarrow{p} \underline{\boldsymbol{\theta}}$

(ii) the distribution of $\sqrt{n}(\hat{\boldsymbol{\theta}}_{QGML} - \underline{\boldsymbol{\theta}})$ *tends to the normal distribution with zero mean vector and covariance matrix* $\boldsymbol{M}_f^{-1} \tilde{\boldsymbol{V}} \boldsymbol{M}_f^{-1}$, *where* $\tilde{\boldsymbol{V}} = \{\tilde{V}_{jl}\}$ *is a* $q \times q$ *matrix such that*

$$\tilde{V}_{jl} = 4\pi \int_{-\pi}^{\pi} \operatorname{tr}\left[f(\lambda) \left\{ \frac{\partial}{\partial \theta_j} f(\lambda)^{-1} \right\} f(\lambda) \left\{ \frac{\partial}{\partial \theta_l} f(\lambda)^{-1} \right\} \right]_{\theta = \underline{\theta}} d\lambda$$

$$+ 2\pi \sum_{r,t,u,v=1}^{m} \iint_{-\pi}^{\pi} \left\{ \frac{\partial}{\partial \theta_j} f_{\theta}^{(r,t)}(\lambda_1) \frac{\partial}{\partial \theta_l} f_{\theta}^{(u,v)}(\lambda_2) \right\}_{\theta = \underline{\theta}} \tag{1.26}$$

$$\times\, Q_{rtuv}^{X}(-\lambda_1, \lambda_2, -\lambda_2) d\lambda_1 d\lambda_2$$

and $f_{\theta}^{(r,t)}(\lambda)$ *is the* (r, t)*th element of* $f_{\theta}(\lambda)^{-1}$.

If $\{X(t)\}$ is Gaussian, then the non-Gaussian quantities satisfy $Q_{rtuv}^{X}(\omega_1, \omega_2, \omega_3) = 0$, hence, the second term in (1.26) vanishes. If $f(\lambda) \equiv f_{\theta}$, i.e., there is no misspecification, then differential calculus for $(\partial/\partial\theta) f_{\theta}^{-1}$ and $(\partial/\partial\theta) \log\det f_{\theta}$ yield,

Corollary 1.3 *Suppose that* $\{X(t)\}$ *is Gaussian, and the spectral density matrix* $f(\lambda) \in \mathcal{F}$, *i.e.,* $f(\lambda) = f_{\theta}(\lambda)$. *Then, under Assumption (HT),*

$$\sqrt{N}(\hat{\theta}_{QGML} - \theta) \xrightarrow{d} N(\mathbf{0}, F(\theta)^{-1}), \tag{1.27}$$

where

$$F(\theta) = \left[\frac{1}{4\pi} \int_{-\pi}^{\pi} \operatorname{tr}\left\{ f_{\theta}(\lambda)^{-1} \left(\frac{\partial}{\partial \theta_j} f_{\theta}(\lambda) \right) f_{\theta}(\lambda)^{-1} \left(\frac{\partial}{\partial \theta_l} f_{\theta}(\lambda) \right) \right\} d\lambda;\ j, l = 1, \cdots, q \right],$$

which is called the (Gaussian) Fisher information matrix in time series.

In view of this corollary, tentatively, we say that an estimator $\hat{\theta}$ of θ is Gaussian asymptotically efficient of $\sqrt{n}(\hat{\theta} - \theta) \xrightarrow{d} N(\mathbf{0}, F(\theta)^{-1})$. The rigorous asymptotic efficiency will discussed in the next section based on the concept of LAN.

We saw in Proposition 1.3 that the asymptotics of $\hat{\theta}_{QGML}$ depend on the integral of the fourth-order cumulant spectra $Q_{rtuv}^{X}(\cdot, \cdot, \cdot)$ which show a degree of non-Gaussianity. We will show that there is a case when the integrals vanish even if the process $\{X(t)\}$ is non-Gaussian.

Assumption (*NGR*)

(i) The innovation process $\{U(t)\}$ in (1.10) satisfies

$$\operatorname{cum}\{u_a(t_1), u_b(t_2), u_c(t_3), u_d(t_4)\} = \begin{cases} \kappa_{abcd} & \text{if } t_1 = t_2 = t_3 = t_4, \\ 0 & \text{otherwise.} \end{cases} \tag{1.28}$$

(ii) $f(\lambda) = f_{\theta}(\lambda)$.

(iii) $\boldsymbol{f}_{\boldsymbol{\theta}}(\lambda)$ has the form $(2\pi)^{-1}\boldsymbol{A}_{\boldsymbol{\theta}}(\lambda)\boldsymbol{K}\boldsymbol{A}_{\boldsymbol{\theta}}(\lambda)^*$, where $\boldsymbol{A}_{\boldsymbol{\theta}}(\lambda) = \sum_{j=0}^{\infty} \boldsymbol{A}_{\boldsymbol{\theta}}(j)e^{-itj}$ with $\boldsymbol{A}_{\boldsymbol{\theta}}(0) = \boldsymbol{I}_m$, and the innovation covariance matrix $\boldsymbol{K}$ is independent of $\boldsymbol{\theta}$, then we say that $\boldsymbol{\theta}$ is "innovation-free."

Evaluating the second integral of (1.26), we can see that all of them vanish, hence,

Corollary 1.4 *Under Assumption (NGR), the asymptotic distribution of $\sqrt{N}(\hat{\boldsymbol{\theta}}_{QGML} - \boldsymbol{\theta})$ is independent of non-Gaussianity of the process (we say that the estimator is non-Gaussian robust (NGR)).*

Our approach due to $D(\boldsymbol{f}_{\boldsymbol{\theta}}, \boldsymbol{f})$ is very wide. For example, we are interested in prediction of $\boldsymbol{X}(t+h)$ based on a linear combination of $\boldsymbol{X}(t), \boldsymbol{X}(t-1), \dots, \boldsymbol{X}(t-r)$, i.e., $\boldsymbol{B}(h)\boldsymbol{X}(t)+\boldsymbol{B}(h+1)\boldsymbol{X}(t-1)+\cdots+\boldsymbol{B}(h+r)\boldsymbol{X}(t-r)$, where $\boldsymbol{B}(h), \dots, \boldsymbol{B}(h+r)$ are $m \times m$ matrices satisfying that $\boldsymbol{I}_m - \boldsymbol{B}(h)z^h - \cdots - \boldsymbol{B}(h+r)z^{h+r}$ is holomorphic on $\mathcal{D} = \{z \in \mathbb{C}; |z| \leq 1\}$.

Let $\boldsymbol{B}_{\boldsymbol{\theta}}(\lambda) = \boldsymbol{I}_m - \boldsymbol{B}(h)e^{ih\lambda} - \cdots - \boldsymbol{B}(h+r)e^{i(h+r)\lambda}$, and $\boldsymbol{f}_{\boldsymbol{\theta}}(\lambda) \equiv \boldsymbol{B}_{\boldsymbol{\theta}}(\lambda)^{-1}\{\boldsymbol{B}_{\boldsymbol{\theta}}(\lambda)^{-1}\}^*$, where $\boldsymbol{\theta} = \text{vec}\{\boldsymbol{B}(h), \dots, \boldsymbol{B}(h+r)\}$. Then the h-step ahead prediction error of $\{\boldsymbol{X}(t+h)\}$ is

$$
\begin{aligned}
\text{tr}E[\{&\boldsymbol{X}(t+h) - \boldsymbol{B}(h)\boldsymbol{X}(t) - \cdots - \boldsymbol{B}(h+r)\boldsymbol{X}(t-r)\} \\
&\times \{\boldsymbol{X}(t+h) - \boldsymbol{B}(h)\boldsymbol{X}(t) - \cdots - \boldsymbol{B}(h+r)\boldsymbol{X}(t-r)\}'] \\
&= \text{tr} \int_{-\pi}^{\pi} \boldsymbol{B}_{\boldsymbol{\theta}}(\lambda)\boldsymbol{f}(\lambda)\boldsymbol{B}_{\boldsymbol{\theta}}(\lambda)^* d\lambda \\
&= \int_{-\pi}^{\pi} \text{tr}\{\boldsymbol{B}_{\boldsymbol{\theta}}(\lambda)^*\boldsymbol{B}_{\boldsymbol{\theta}}(\lambda)\boldsymbol{f}(\lambda)\}d\lambda \\
&= \int_{-\pi}^{\pi} \text{tr}\{\boldsymbol{f}_{\boldsymbol{\theta}}{}^{-1}(\lambda)\boldsymbol{f}(\lambda)\}d\lambda
\end{aligned} \tag{1.29}
$$

From Theorem 3''' of Hannan (1970), p.162 it follows that

$$
\int_{-\pi}^{\pi} \log\det \boldsymbol{f}_{\boldsymbol{\theta}}(\lambda)d\lambda = 0 \tag{1.30}
$$

which, together with (1.29), implies that the h-step ahead prediction error is equal to $D(\boldsymbol{f}_{\boldsymbol{\theta}}, \boldsymbol{f})$. Let $\underline{\boldsymbol{\theta}}$ be the quasi-true value of $\boldsymbol{\theta}$, i.e.,

$$
\underline{\boldsymbol{\theta}} = \text{vec}\{\underline{\boldsymbol{B}}(h), \dots, \underline{\boldsymbol{B}}(h+r)\} \equiv \arg\min_{\boldsymbol{\theta}\in\boldsymbol{\Theta}} D(\boldsymbol{f}_{\boldsymbol{\theta}}, \boldsymbol{f}).
$$

Then, the corresponding predictor

$$\underline{\boldsymbol{B}}(h)\boldsymbol{X}(t) + \cdots + \underline{\boldsymbol{B}}(h+r)\boldsymbol{X}(t-r) \tag{1.31}$$

becomes the best h-step ahead linear predictor on $\{\boldsymbol{X}(t), \ldots, \boldsymbol{X}(t-r)\}$. Observing the data $\{\boldsymbol{X}(t), \boldsymbol{X}(t-1), \ldots, \boldsymbol{X}(t-n+1)\}$, we can estimate $\underline{\boldsymbol{\theta}}$ by

$$\hat{\boldsymbol{\theta}}_{QGML} \equiv \text{vec}\{\hat{\underline{\boldsymbol{B}}}_{QGML}(h), \ldots, \hat{\underline{\boldsymbol{B}}}_{QGML}(h+r)\} \tag{1.32}$$

whose asymptotics are given in Proposition 1.3 with $\dim \boldsymbol{\theta} = m(r+1)$, i.e.,

$$\sqrt{N}(\hat{\boldsymbol{\theta}}_{QGML} - \boldsymbol{\theta}) \xrightarrow{d} N(\boldsymbol{0}, \boldsymbol{M}_f^{-1}\tilde{\boldsymbol{V}}\boldsymbol{M}_f^{-1}). \tag{1.33}$$

Let us consider the problem of portfolio. Suppose that we have m-asset returns described by $X_1(t), \ldots, X_m(t)$. Let $\boldsymbol{\alpha} = (\alpha_1, \ldots, \alpha_m)'$ be a portfolio weight vector satisfying $\alpha_1 + \cdots + \alpha_m = 1$. Assuming that the present time is t, we want to estimate the future portfolio return $\boldsymbol{\alpha}'\boldsymbol{X}(t+h)$. In view of above, this is estimated by

$$\boldsymbol{\alpha}'\left\{\hat{\underline{\boldsymbol{B}}}_{QGML}(h)\boldsymbol{X}(t) + \cdots + \hat{\underline{\boldsymbol{B}}}_{QGML}(h+r)\boldsymbol{X}(t-r)\right\}. \tag{1.34}$$

Hence our setting by $D(\boldsymbol{f}_{\boldsymbol{\theta}}, \boldsymbol{f})$ is unexpectedly general and wide.

We next discuss the problem of nonparametric spectral estimation. We estimate $\boldsymbol{f}(\lambda)$ by weighted averages of the periodogram matrix $\boldsymbol{I}_n(\lambda)$ with a spectral window $W_n(\lambda)$ as weight, i.e.,

$$\hat{\boldsymbol{f}}_n(\lambda) = \int_{-\pi}^{\pi} W_n(\lambda - \mu)\boldsymbol{I}_n(\mu)d\mu \tag{1.35}$$

where $W_n(\cdot)$ satisfies

Assumption 1.2

(i) $W_n(\lambda)$ can be expanded as

$$W_n(\lambda) = \frac{1}{2\pi}\sum_{l=-M}^{M} w\left(\frac{l}{M}\right)\exp(-il\lambda).$$

(ii) $w(x)$ is continuous, even function with $w(0) = 1$, and satisfies

$$\begin{cases} |w(x)| \le 1, \\ \int_{-\infty}^{\infty} w(x)^2 dx < \infty, \\ \lim_{x\to 0} \dfrac{1-w(x)}{|x|^2} = \kappa_2 < \infty. \end{cases}$$

(iii) $M = M(n)$ satisfies

$$n^{1/4}/M + M/n^{1/2} \to 0 \text{ as } n \to \infty.$$

Concrete examples of $W_n(\cdot)$ satisfying Assumption 1.2 are found in e.g., Hannan (1970), and it is known that

$$E\left[||\hat{\boldsymbol{f}}_n(\lambda) - \boldsymbol{f}(\lambda)||^2\right] = \left(\frac{M}{n}\right) + O(M^{-4}) \tag{1.36}$$

uniformly in λ, hence, $\hat{\boldsymbol{f}}_n(\lambda) - \boldsymbol{f}(\lambda) = O_p\{(M/n)^{1/2}\}$, i.e., $\hat{\boldsymbol{f}}(\lambda)$ is a $\sqrt{n/M}$-consistent estimator of $\boldsymbol{f}(\lambda)$ (e.g., Hannan (1970)). As we saw in Proposition 1.3, parametric estimators usually have $\sqrt{n}$-consistency, so the result above means a disadvantage of nonparametric spectral estimators. But, in what follows, we will show that this disadvantage disappears if we consider appropriate integral functionals of $\hat{\boldsymbol{f}}_n(\lambda)$.

Let D be an open subset of $\mathbb{C}^{m^2}$.

Assumption 1.3 A mapping $K : D \to \mathbb{R}$ is holomorphic.

The integral functional of $\boldsymbol{f}(\lambda)$

$$\int_{-\pi}^{\pi} K\{\boldsymbol{f}(\lambda)\} d\lambda \tag{1.37}$$

can represent a wide variety of important time series indices.

The example in financial econometrics will be given later. If we are interested in estimation of (1.37), it is very natural to use $\int_{-\pi}^{\pi} K\{\hat{\boldsymbol{f}}_n(\lambda)\}d\lambda$ as an estimator of (1.37). Denote by $\boldsymbol{K}^{(1)}\{\boldsymbol{f}(\lambda)\} = [K_{ab}^{(1)}\{\boldsymbol{f}(\lambda)\}; a, b = 1, \ldots, m]$ the first-order derivative of $K\{\boldsymbol{f}(\lambda)\}$ at $\boldsymbol{f}(\lambda)$ (see Magnus and Neudecker (1988)). Showing the relation

$$\sqrt{n}\left[\int_{-\pi}^{\pi} K\{\hat{\boldsymbol{f}}(\lambda)\}d\lambda - \int_{-\pi}^{\pi} K\{\boldsymbol{f}(\lambda)\}d\lambda\right] = \sqrt{n}\int_{-\pi}^{\pi}\left[\operatorname{tr}\{\boldsymbol{I}_n(\lambda) - \boldsymbol{f}(\lambda)\}\boldsymbol{K}^{(1)}\{\boldsymbol{f}(\lambda)\}\right]d\lambda + o_p(1), \tag{1.38}$$

Taniguchi et al. (1996) gave the following.

Proposition 1.4 *Suppose that Assumption (B) or (HT) holds. Then, under Assumptions 1.2 and 1.3, as $n \to \infty$,*

$$\sqrt{n}\left[\int_{-\pi}^{\pi} K\{\hat{\boldsymbol{f}}(\lambda)\}d\lambda - \int_{-\pi}^{\pi} K\{\boldsymbol{f}(\lambda)\}d\lambda\right] \xrightarrow{d} N\left(0, \nu_1(\boldsymbol{f}) + \nu_2(Q^{\boldsymbol{X}})\right), \tag{1.39}$$

where

$$\nu_1(\boldsymbol{f}) = 4\pi \int_{-\pi}^{\pi} \mathrm{tr}\left[\boldsymbol{f}(\lambda)\boldsymbol{K}^{(1)}\{\boldsymbol{f}(\lambda)\}\right]^2 d\lambda, \tag{1.40}$$

and

$$\nu_2(Q^{\boldsymbol{X}}) = 2\pi \sum_{a,b,c,d=1}^{m} \iint_{-\pi}^{\pi} K_{ab}^{(1)}\{\boldsymbol{f}(\lambda)\}K_{cd}^{(1)}\{\boldsymbol{f}(\lambda)\}Q_{abcd}^{\boldsymbol{X}}(-\lambda_1, \lambda_2, -\lambda_2)d\lambda_1 d\lambda_2. \tag{1.41}$$

It may be noted that $\sqrt{n}$-consistency holds in Proposition 1.4 despite $\hat{\boldsymbol{f}}_n(\lambda)$ being a $\sqrt{n/M}$-consistent estimator of $\boldsymbol{f}(\lambda)$. This is due to the fact that integration of $\hat{\boldsymbol{f}}_n(\lambda)$ recovers $\sqrt{n}$-consistency.

Consider the problem of testing

$$\begin{aligned} \mathrm{H} &: \int_{-\pi}^{\pi} K\{\boldsymbol{f}(\lambda)\}d\lambda = c \\ & \qquad\qquad against \\ \mathrm{A} &: \int_{-\pi}^{\pi} K\{\boldsymbol{f}(\lambda)\}d\lambda \neq c, \end{aligned} \tag{1.42}$$

where c is a given constant. Under H, it follows from Proposition 1.4 that

$$S_n \equiv \frac{\sqrt{n}[\int_{-\pi}^{\pi} K\{\hat{\boldsymbol{f}}_n(\lambda)\}d\lambda - c]}{\sqrt{\nu_1(\boldsymbol{f}) + \nu_2(Q^{\boldsymbol{X}})}} \xrightarrow{d} N(0, 1), \tag{1.43}$$

as $n \to \infty$. However, we have to estimate the denominator if S_n is feasible for testing (1.42). Regarding $\nu_1(\boldsymbol{f})$, $\nu_1(\hat{\boldsymbol{f}}_n)$ becomes a consistent estimator of it. Taniguchi (1982) gave a consistent estimator for quantities of the form of $\nu_2(Q^{\boldsymbol{X}})$, hence, we write this estimator as $\widehat{\nu_2(Q^{\boldsymbol{X}})}$. Consequently, it is seen that, as $n \to \infty$,

$$T_n \equiv \frac{\sqrt{n}[\int_{-\pi}^{\pi} K\{\hat{f}_n(\lambda)\}d\lambda - c]}{\sqrt{\nu_1(\hat{f}_n) + \widehat{\nu_2(Q^X)}}} \xrightarrow{d} N(0,1), \tag{1.44}$$

under H. We can use T_n as a test statistic for (1.42). It may be noted that T_n is essentially nonparametric, and has $\sqrt{n}$-consistency.

Because our integral functional $\int_{-\pi}^{\pi} K\{f(\lambda)\}d\lambda$ can represent so many important indices in time series, the test setting in (1.42) is very wide. In what follows, a few examples are provided. Suppose the process $\{X(t) : t \in \mathbb{Z}\}$ is of the form

$$X(t) = \begin{bmatrix} x(t) \\ y(t) \end{bmatrix}$$

with $x(t)$, q vector-valued, and $y(t)$, r vetor-valued; $q + r = m$ and has the spectral density matrix

$$f(\lambda) = \begin{bmatrix} f_{xx}(\lambda) & f_{xy}(\lambda) \\ f_{yx}(\lambda) & f_{yy}(\lambda) \end{bmatrix}. \tag{1.45}$$

We denote by $H\{\cdot\}$ the linear closed manifold generated by $\{\cdot\}$, and denote by $\mathcal{P}[x(t)|H\{\cdot\}]$ the linear projection of $x(t)$ on $H\{\cdot\}$.

Consider the residual process

$$\begin{aligned}
u_1(t) &= x(t) - \mathcal{P}[x(t)|H\{x(t-1), x(t-2), \ldots\}], \\
v_1(t) &= y(t) - \mathcal{P}[y(t)|H\{y(t-1), y(t-2), \ldots\}], \\
u_2(t) &= x(t) - \mathcal{P}[x(t)|H\{x(t-1), x(t-2), \ldots; y(t-1), y(t-2), \ldots\}], \\
v_2(t) &= y(t) - \mathcal{P}[y(t)|H\{x(t-1), x(t-2), \ldots; y(t-1), y(t-2), \ldots\}],
\end{aligned}$$

and

$$u_3(t) = x(t) - \mathcal{P}[x(t)|H\{x(t-1), x(t-2), \ldots; y(t), y(t-1), \ldots\}].$$

The measure of linear feedback from $Y = \{y(t)\}$ to $X = \{x(t)\}$ is defined by

$$F_{Y \to X} = \log[\det\{Var(u_1(t))\}/\det\{Var(u_2(t))\}] \tag{1.46}$$

Symmetrically, we can define

$$F_{X \to Y} = \log[\det\{Var(v_1(t))\}/\det\{Var(v_2(t))\}] \tag{1.47}$$

The measure of instantaneous linear feedback

$$F_{X \cdot Y} = \log[\det\{Var(u_2(t))\}/\det\{Var(u_3(t))\}]$$

has motivation similar to that of the above two measures. The following

$$F_{X,Y} = \log[\det\{Var(u_1(t))\}\det\{Var(v_1(t))\}/\det\{Var(U(t))\}]$$

is called the measure of linear dependence. Then it is shown that

$$F_{X,Y} = F_{Y \to X} + F_{X \to Y} + F_{X \cdot Y} \tag{1.48}$$

and

$$F_{X \cdot Y} = \int_{-\pi}^{\pi} K\{\boldsymbol{f}(\lambda)\} d\lambda, \tag{1.49}$$

where

$$K\{\boldsymbol{f}(\lambda)\} = -\frac{1}{2\pi} \log[\det\{\boldsymbol{I}_q - \boldsymbol{f}_{xy}(\lambda)\boldsymbol{f}_{yy}(\lambda)^{-1}\boldsymbol{f}_{yx}(\lambda)\boldsymbol{f}_{xx}(\lambda)^{-1}\}],$$

(see Geweke (1982).) Because $F_{Y \to X}$, $F_{X \to Y}$ and $F_{X \cdot Y}$ are important econometric measures which represent "strength of causality," the problem

$$H : F_{X,Y} = c, \quad \text{against} \quad A : F_{X,Y} \neq c, \tag{1.50}$$

is exactly an example of our testing (1.42). Therefore, we can test (1.50) by use of T_n given in (1.44).

Next we discuss another interrelation analysis. Let $\boldsymbol{X}(t) = \{\boldsymbol{x}(t)', \boldsymbol{y}(t)', \boldsymbol{z}(t)'\}'$ be the m-dimensional linear process given in (1.10), where $\boldsymbol{x}(t)$, $\boldsymbol{y}(t)$ and $\boldsymbol{z}(t)$ are q, r and s $(q + r + s = m)$ component processes, respectively. Correspondingly we write the spectral density matrix and the spectral representation, respectively, as

$$\boldsymbol{f}(\lambda) = \begin{bmatrix} \boldsymbol{f}_{xx}(\lambda) & \boldsymbol{f}_{xy}(\lambda) & \boldsymbol{f}_{xz}(\lambda) \\ \boldsymbol{f}_{yx}(\lambda) & \boldsymbol{f}_{yy}(\lambda) & \boldsymbol{f}_{yz}(\lambda) \\ \boldsymbol{f}_{zx}(\lambda) & \boldsymbol{f}_{zy}(\lambda) & \boldsymbol{f}_{zz}(\lambda) \end{bmatrix},$$

and

$$\boldsymbol{X}(t) = \int_{-\pi}^{\pi} e^{-it\lambda} d\boldsymbol{\xi}(\lambda),$$

where $d\boldsymbol{\xi}(\lambda) = (d\boldsymbol{\xi}_x(\lambda)', d\boldsymbol{\xi}_y(\lambda)', d\boldsymbol{\xi}_z(\lambda)')'$. Hannan (1970) considered a test for association for $\boldsymbol{x} = \{\boldsymbol{x}(t)\}$ with $\boldsymbol{y} = \{\boldsymbol{y}(t)\}$ (at frequency λ) after allowing for any affects of $\boldsymbol{z} = \{\boldsymbol{z}(t)\}$. The hypothesis is given by

$$\mathrm{H}_\lambda : \boldsymbol{f}_{xy}(\lambda) - \boldsymbol{f}_{xz}(\lambda)\boldsymbol{f}_{zz}(\lambda)^{-1}\boldsymbol{f}_{zy}(\lambda) = \boldsymbol{0}, \tag{1.51}$$

which means that $d\boldsymbol{\xi}_x(\lambda) - \boldsymbol{f}_{xz}(\lambda)\boldsymbol{f}_{zz}(\lambda)^{-1}d\boldsymbol{\xi}_z(\lambda)$ is incoherent with $d\boldsymbol{\xi}_y(\lambda) - \boldsymbol{f}_{yz}(\lambda)\boldsymbol{f}_{zz}(\lambda)^{-1}d\boldsymbol{\xi}_z(\lambda)$, and all of the apparent association between $\boldsymbol{x}$ and $\boldsymbol{y}$ is truly due only to their common association with $\boldsymbol{z}$. For a given λ, Hannan (1970) developed the testing theory for H_λ based on the asymptotic normality of the finite

Fourier transformations of $\{\boldsymbol{X}(t)\}$ in a neighborhood of λ. In the context of (1.42), we can introduce a test for association for $\boldsymbol{x}$ and $\boldsymbol{y}$ at "all the frequency $\lambda \in [-\pi, \pi]$" after allowing for any effects z. The hypothesis is written as

$$\mathrm{H} : \boldsymbol{f}_{\boldsymbol{xy}}(\lambda) - \boldsymbol{f}_{\boldsymbol{xz}}(\lambda)\boldsymbol{f}_{\boldsymbol{zz}}(\lambda)^{-1}\boldsymbol{f}_{\boldsymbol{xy}}(\lambda) = \boldsymbol{0} \quad \textit{for all } \lambda \in [-\pi, \pi],$$

equivalently,

$$\mathrm{H} : \int_{-\pi}^{\pi} K\{\boldsymbol{f}(\lambda)\} d\lambda = 0, \tag{1.52}$$

where

$$K\{\boldsymbol{f}(\lambda)\} = \mathrm{tr}[\{\boldsymbol{f}_{\boldsymbol{xy}}(\lambda) - \boldsymbol{f}_{\boldsymbol{xz}}(\lambda)\boldsymbol{f}_{\boldsymbol{zz}}(\lambda)^{-1}\boldsymbol{f}_{\boldsymbol{zy}}(\lambda)\}\{\boldsymbol{f}_{\boldsymbol{yx}}(\lambda) - \boldsymbol{f}_{\boldsymbol{yz}}(\lambda)\boldsymbol{f}_{\boldsymbol{zz}}(\lambda)^{-1}\boldsymbol{f}_{\boldsymbol{zx}}(\lambda)\}].$$

Then, we can test (1.52) by use of T_n, hence, we can grasp Hannan's problem above as ours.

Now, let us think of estimation based on integral functional of $\hat{\boldsymbol{f}}_n$. Kakizawa (1996) introduced the following generalized disparity measure between spectral density matrices $\boldsymbol{f}(\lambda)$ and $\boldsymbol{g}(\lambda)$;

$$D_H(\boldsymbol{f} : \boldsymbol{g}) = \frac{1}{4\pi} \int_{-\pi}^{\pi} H\{\boldsymbol{f}(\lambda)\boldsymbol{g}(\lambda)^{-1}\} d\lambda, \tag{1.53}$$

where $H(\boldsymbol{Z})$ is a holomorphic function of $m \times m$ matrix $\boldsymbol{Z}$, and has a unique minimum zero at $\boldsymbol{Z} = \boldsymbol{I}_n$, such as

$$\begin{aligned}
H_1(\boldsymbol{Z}) &= -\log\det(\boldsymbol{Z}) + \mathrm{tr}(\boldsymbol{Z}) - m,\\
H_2(\boldsymbol{Z}) &= \frac{1}{2}\mathrm{tr}(\boldsymbol{Z} - \boldsymbol{I}_m)^2,\\
H_3(\boldsymbol{Z}) &= \frac{1}{\alpha(1-\alpha)}\left[\log\det\{(1-\alpha)\boldsymbol{I}_m + \alpha\boldsymbol{Z}\} - \alpha\log\det\boldsymbol{Z}\right], \quad \alpha \in (0, 1).
\end{aligned}$$

In fact, $H(\boldsymbol{Z})$ is defined by

$$H(\boldsymbol{Z}) = \sum_{l \neq 0}\sum_{j=1}^{3} p_{j,l}\frac{H_j(\boldsymbol{Z}^l)}{l^2}, \tag{1.54}$$

where $p_{j,l} \geq 0$ for $j = 1, 2, 3; l = \pm 1, \pm 2, \ldots$ and $\sum_l \sum_{j=1}^{3} p_{j,l} = 1$. Thus $D_H(\boldsymbol{f} : \boldsymbol{g})$ includes many other criteria as special cases. Let

$$\hat{\boldsymbol{\theta}}_H = \arg\min_{\boldsymbol{\theta} \in \boldsymbol{\Theta}} D_H(\boldsymbol{f}_{\boldsymbol{\theta}}, \hat{\boldsymbol{f}}_n) \tag{1.55}$$

The following proposition is due to Kakizawa (1996) and Taniguchi and Kakizawa (2000).

Proposition 1.5 *Suppose all the assumptions in Proposition 1.3 hold, and the true spectral density matrix is of the form $\boldsymbol{f}_{\boldsymbol{\theta}}$. Then, under Gaussianity of the process,*

$$\sqrt{n}(\hat{\boldsymbol{\theta}}_H - \boldsymbol{\theta}) \xrightarrow{d} N(\mathbf{0}, \boldsymbol{F}(\boldsymbol{\theta})^{-1}), \tag{1.56}$$

as $n \to \infty$, i.e., $\hat{\boldsymbol{\theta}}_H$ has the same asymptotics as $\hat{\boldsymbol{\theta}}_{QGML}$ does (see Corollary 1.3).

Proposition 1.5 implies that we can construct infinitely many Gaussian asymptotically efficient estimators. In Sect. 1.3 we will show that $\hat{\boldsymbol{\theta}}_H$ has a robustness which $\hat{\boldsymbol{\theta}}_{QGML}$ does not share.

1.3 Optimal Inference for Various Return Processes

As we saw in the previous subsection, there are a lot of time series models (e.g., ARMA, linear process, ARCH, GARCH, EGARCH, SV, CHARN, etc.). Here, we develop the asymptotic optimal estimation theory based on local asymptotic normality (LAN) for typical time series models.

Lucien Le Cam established one of the most important foundations of the general statistical asymptotic theory (e.g., see LeCam (1986)). He introduced the concept of LAN for the likelihood ratio of general statistical model. Once LAN is proved, the asymptotic optimality of estimators and tests is described in terms of the LAN property.

We briefly review the LAN results for stochastic processes. Swensen (1985) showed that the likelihood ratio of an autoregressive process of finite order with a trend is LAN. Regarding ARMA processes, Hallin et al. (1985) and Kreiss (1987) showed the LAN property, and applied the results to test and estimation theory. Further, Kreiss (1990) developed the LAN theory for a class of autoregressive processes with infinite order. Garel and Hallin (1995) proved the LAN for multivariate general linear models with ARMA residual.

As an important class of nonlinear time series models, Giraitis et al. (2000) introduced a class of ARCH(∞) models, which includes the ARCH and GARCH models as special cases, and gave sufficient conditions for the existence of a stationary solution and its explicit representation.

Here we deal with the more general ARCH(∞) model with stochastic mean (ARCH(∞)-SM model). In what follows we provide the LAN results and asymptotic optimal estimation for ARCH(∞)-SM model in line with Lee and Taniguchi (2005).

Suppose that $(\Omega, \mathcal{F}, P)$ is a probability space, and $\{\mathcal{F}_t : t \in \mathbb{Z}\}$ is a sequence of sub-σ-algebras of $\mathcal{F}$ satisfying $\mathcal{F}_t \subset \mathcal{F}_{t+1}$, $t \in \mathbb{Z}$. Consider the ARCH(∞)-SM model $\{Y_t : t \in \mathbb{Z}\}$ defined by

$$\begin{cases} Y_t - \boldsymbol{\beta}' z_t = \sigma_t u_t, & t \in \mathbb{Z}, \\ \sigma_t^2 = a + \sum_{j=1}^{\infty} b_j (Y_{t-j} - \boldsymbol{\beta}' z_{t-j})^2. \end{cases} \tag{1.57}$$

where $a > 0$, $b_j \geq 0$, $j = 1, 2, \ldots$, $\{u_t : t \in \mathbb{Z}\}$ is a sequence of i.i.d. random variables with density $g(\cdot)$, and u_t is $\mathcal{F}_t$-measurable and independent of $\mathcal{F}_{t-1}$. Here $\boldsymbol{\beta} = (\beta_1, \cdots, \beta_p)'$ is an unknown vector and the $z_t = (z_{t_1}, \ldots, z_{t_p})'$s are observable $p \times 1$ random vectors which are $\mathcal{F}_{t-1}$-measurable. If $\boldsymbol{\beta} = \mathbf{0}$, this model reduces to the ARCH(∞) model proposed by Giraitis et al. (2000). We can see that the class of ARCH(∞)-SM models is larger than that of GARCH(r, s)-SM models defined by

$$\begin{cases} Y_t - \boldsymbol{\beta}' z_t = \sigma_t u_t, \\ \sigma_t^2 = a + \sum_{j=1}^{r} a_j \sigma_{t-j}^2 + \sum_{j=1}^{s} b_j (Y_{t-j} - \boldsymbol{\beta}' z_{t-j})^2. \end{cases} \tag{1.58}$$

where the associated polynomials of the second equation of (1.58) satisfy the invertible condition. If we take $z_t = (Y_{t-1}, Y_{t-2}, \ldots, Y_{t-p})'$, then (1.58) becomes the AR(p)-GARCH(r, s) models, which implies that the class of ARCH(∞)-SM models is sufficiently extensive. To develop the asymptotic theory for (1.58) we impose,

Assumption 1.4

(i) $E\{u_t\} = 0$, $\text{Var}\{u_t\} = 1$ and $E\{u_t^4\} < \infty$

(ii) a and b_j's are function of an unknown parameter $\boldsymbol{\eta} = (\eta_1, \ldots, \eta_q)'$, i.e., $a = a(\boldsymbol{\eta})$ and $b_j = b_j(\boldsymbol{\eta})$ for $j \geq 1$ and $\boldsymbol{\eta} \in \mathcal{H}$, where $\mathcal{H}$ is an open subset of $\mathbb{R}^q$. The functions $a(\boldsymbol{\eta})$ and $b_j(\boldsymbol{\eta})$ are twice continuously differentiable with respect to $\boldsymbol{\eta}$.

(iii) There exist $\tilde{a} > 0$ and $\tilde{b}_j \geq 0$ satisfying $\sum_{j=1}^{\infty} \tilde{b}_j < 1$, such that $a(\boldsymbol{\eta}) \geq \tilde{a}$ and $b_j(\boldsymbol{\eta}) \leq \tilde{b}_j$ for all $j \geq 1$ and $\boldsymbol{\eta} \in \mathcal{H}$, which entails

$$\sum_{j=1}^{\infty} b_j(\boldsymbol{\eta}) < 1 \quad for\ all\ \boldsymbol{\eta} \in \boldsymbol{\mathcal{H}}. \tag{1.59}$$

(iv) There exist $\tilde{a}^{(1)} > 0, \tilde{a}^{(2)} > 0$ and $\tilde{b}_j^{(1)} \geq 0, \tilde{b}_j^{(2)} \geq 0$ satisfying $\sum_{j=1}^{\infty} \tilde{b}_j^{(i)} < \infty$, $i = 1, 2$, such that $||\partial a(\boldsymbol{\eta})/\partial \boldsymbol{\eta}|| \leq \tilde{a}^{(1)}$, $||(\partial^2/\partial \boldsymbol{\eta} \partial \boldsymbol{\eta}') a(\boldsymbol{\eta})|| \leq \tilde{a}^{(2)}$ for all $\boldsymbol{\eta} \in \mathcal{H}$, and $||(\partial/\partial \boldsymbol{\eta}) b_j(\boldsymbol{\eta})|| \leq \tilde{b}_j^{(1)}$, $||(\partial^2/\partial \boldsymbol{\eta} \partial \boldsymbol{\eta}') b_j(\boldsymbol{\eta})|| \leq \tilde{b}^{(2)}$ for all $j \geq 1$ and $\boldsymbol{\eta} \in \mathcal{H}$, where $||\boldsymbol{a}||$ denotes the Euclidean norm of a vector or matrix $\boldsymbol{a}$, i.e., $\sqrt{\text{tr}(\boldsymbol{a}'\boldsymbol{a})}$.

The condition (1.59) guarantees the existence of strictly stationary solution for $\{Y_t - \boldsymbol{\beta}' z_t\}$ in (1.58) (c.f. Giraitis et al. (2000)).

Assumption 1.5 $\{E(u_t^4)\}^{1/2} \sum_{j=1}^{\infty} b_j < 1$ and $E||z_t||^4 < \infty$.

The condition $\{E(u_t^4)\}^{1/2} \sum_{j=1}^{\infty} b_j < 1$ implies $E(Y_t^4) < \infty$ (see Giraitis et al. (2000)). In a special case of GARCH models in (1.58), a necessary and sufficient condition for the existence of the fourth moment of $Y_t - \boldsymbol{\beta}' z_t$ was established by Ling and Li (1997), Chen and An (1998) and Ling and McAleer (2003).

Assumption 1.6 The innovation density $g(\cdot)$ is symmetric, twice continuously differentiable, and satisfies

(i) $0 < I(g) \equiv \int_{-\infty}^{\infty} \{g'(u)/g(u)\}^2 g(u)du < \infty$, and $\int_{-\infty}^{\infty} \{g'(u)/g(u)\}^4 g(u)du < \infty$.
(ii) $\lim_{|u|\to\infty} ug(u) = 0$, $\lim_{|u|\to\infty} u^2 g'(u) = 0$.

Assumption 1.7 The matrix $n^{-1}\sum_{t=1}^{n} z_t z_t'/\sigma_t^2$ converges to a finite limit $\boldsymbol{M}(0)$ in L^2-sense, where $\boldsymbol{M}(0)$ is positive definite.

We write $\boldsymbol{\theta} = (\boldsymbol{\beta}', \boldsymbol{\eta}')' \in \boldsymbol{\Theta}$ and $\dim \boldsymbol{\Theta} = r = p + q$, where $\boldsymbol{\Theta}$ is an open subset of $\mathbb{R}^r$. Then the $\sigma_t = \sigma_t(\boldsymbol{\theta})$'s are measurable functions of $\boldsymbol{\theta}$ and Y_{t-j}, $j \geq 1$. Let $P_{\boldsymbol{\theta}}^{(n)}$ be the distribution of $(u_s, s \leq 0, Y_1, \ldots, Y_n)$. For two hypothetical values $\boldsymbol{\theta}, \boldsymbol{\theta}' \in \boldsymbol{\Theta}$ the log-likelihood ratio is written as

$$\Lambda_n(\boldsymbol{\theta}, \boldsymbol{\theta}') \equiv \log \frac{dP_{\boldsymbol{\theta}'}^{(n)}}{dP_{\boldsymbol{\theta}}^{(n)}} = 2\sum_{t=1}^{n} \log \Phi_t^{(n)}(\boldsymbol{\theta}, \boldsymbol{\theta}'), \qquad (1.60)$$

where $\Phi_t^{(n)}(\boldsymbol{\theta}, \boldsymbol{\theta}') = \left[\{g\{\phi_t(\boldsymbol{\theta}')\}\sigma_t(\boldsymbol{\theta})\} / \{g\{\phi_t(\boldsymbol{\theta})\}\sigma_t(\boldsymbol{\theta}')\}\right]^{1/2}$ with $\phi_t(\boldsymbol{\theta}) = (Y_t - \boldsymbol{\beta}' z_t)/\sigma_t(\boldsymbol{\theta})$. We denote by $H(g; \boldsymbol{\theta})$ the hypothesis under which the underlying parameter is $\boldsymbol{\theta} \in \boldsymbol{\Theta}$ and the density of u_t is $g = g(\cdot)$. Define

$$\boldsymbol{\theta}_n = (\boldsymbol{\beta}_n', \boldsymbol{\eta}_n')' \equiv \boldsymbol{\theta} + \frac{1}{\sqrt{n}}\boldsymbol{\xi}, \quad \boldsymbol{\xi} = (\boldsymbol{\kappa}', \boldsymbol{h}')' \in S \subset \mathbb{R}^r,$$

where $\boldsymbol{\kappa} = (\kappa_1, \ldots, \kappa_p)'$, $\boldsymbol{h} = (h_1, \ldots, h_q)'$, and S is an open subset of $\mathbb{R}^r$. In what follows we denote by $\mathbb{R}^{\mathbb{Z}}$ the product space $\cdots \times \mathbb{R} \times \mathbb{R} \times \mathbb{R} \times \mathbb{R} \times \cdots$, whose component spaces correspond to the coordinate spaces of $(\ldots, u_{-1}, u_0, Y_1, Y_2, \ldots)$, and write its Borel σ-algebra by $\mathcal{B}^{\mathbb{Z}}$.

The following proposition is due to Lee and Taniguchi (2005).

Proposition 1.6 (LAN for ARCH(∞)-SM models) *Suppose that Assumptions 1.4–1.7 hold. Then the sequence of experiments* $\mathcal{E}_n = \{\mathbb{R}^{\mathbb{Z}}, \mathcal{B}^{\mathbb{Z}}, \{P_{\boldsymbol{\theta}}^{(n)} : \boldsymbol{\theta} \in \boldsymbol{\Theta} \subset \mathbb{R}^r\}\}$, $n \in \mathbb{N}$, *is locally asymptotically normal and equicontinuous on compact subset C of S. That is,*

(i) For all $\boldsymbol{\theta} \in \boldsymbol{\Theta}$, the log-likelihood ratio $\Lambda_n(\boldsymbol{\theta}, \boldsymbol{\theta}_n)$ admits the following stochastic expansion under $H(g; \boldsymbol{\theta})$:

$$\Lambda_n(\boldsymbol{\theta}, \boldsymbol{\theta}_n) = (\boldsymbol{\kappa}', \boldsymbol{h}')\frac{1}{\sqrt{n}}\sum_{t=1}^{n}(\boldsymbol{\Delta}_{1,t}', \boldsymbol{\Delta}_{2,t}')' - \frac{1}{2}\boldsymbol{\xi}' \boldsymbol{F} \boldsymbol{\xi} + o_p(1), \qquad (1.61)$$

where $\boldsymbol{\Delta}_{1,t} = -(z_t/\sigma_t)(g'(\phi_t)/g(\phi_t)) - (2\sigma_t^2)^{-1}(\partial\sigma_t^2/\partial\boldsymbol{\beta})\{1 + \phi_t(g'(\phi_t)/g(\phi_t))\}$, $\boldsymbol{\Delta}_{2,t} = -(2\sigma_t^2)^{-1}\{1 + \phi_t(g'(\phi_t)/g(\phi_t))\}\,(\partial/\partial\boldsymbol{\eta})\sigma_t^2$, *and*

$$\boldsymbol{F} = \begin{pmatrix} \boldsymbol{F}_{11}, & \boldsymbol{F}_{12} \\ \boldsymbol{F}_{21}, & \boldsymbol{F}_{22} \end{pmatrix}.$$

Here

$$\begin{aligned}
\boldsymbol{F}_{11} &= I(g)\cdot \boldsymbol{M}(0) + \{J(g) - 1\}\, E[(4\sigma_t^4)^{-1}(\partial\sigma_t^2/\partial\boldsymbol{\beta})(\partial\sigma_t^2/\partial\boldsymbol{\beta}')],\\
\boldsymbol{F}_{12} &= \{J(g) - 1\}\, E[(4\sigma_t^4)^{-1}(\partial\sigma_t^2/\partial\boldsymbol{\beta})(\partial\sigma_t^2/\partial\boldsymbol{\eta}')],\\
\boldsymbol{F}_{22} &= \{J(g) - 1\}\, E[(4\sigma_t^4)^{-1}(\partial\sigma_t^2/\partial\boldsymbol{\eta})(\partial\sigma_t^2/\partial\boldsymbol{\eta}')],
\end{aligned}$$

where $J(g) = E\{u_t^2(g'(u_t)/g(u_t))^2\}$.

(ii) *Under* $H(g;\boldsymbol{\theta})$, $\boldsymbol{\Delta}_n \xrightarrow{d} N(\boldsymbol{0}, \boldsymbol{F})$, *where* $\boldsymbol{\Delta}_n = n^{-1/2}\sum_{t=1}^{2}(\boldsymbol{\Delta}_{1,t}', \boldsymbol{\Delta}_{2,t}')'$.

(iii) *For all* $n \in \mathbb{N}$ *and* $\boldsymbol{\xi} \in S$, *the mapping* $\boldsymbol{\xi} \to P_{\boldsymbol{\theta}_n}^{(n)}$ *is continuous with respect to the variational distance* $||P - Q|| = \sup\{|P(A) - Q(A)|;\ A \in B^{\mathbb{Z}}\}$.

The term $\boldsymbol{\Delta}_n$, called the central sequence, is measurable with respect to u_s, $s \leq 0$, Y_j, z_j, $j = 1, \ldots, n$, but it is not so with respect to the the observable sequences, Y_j, z_j, $j = 1, \ldots, n$. Therefore, we will construct a $(Y_1, \ldots, Y_n, z_1, \ldots, z_n)$-measurable version $\tilde{\boldsymbol{\Delta}}_n$ of $\boldsymbol{\Delta}_n$. For this, we introduce the truncated versions of σ_t^2 and ϕ_t by $\tilde{\sigma}_t^2 = \tilde{\sigma}_t^2(\boldsymbol{\beta}, \boldsymbol{\eta}) \equiv a + \sum_{j=1}^{t-1} b_j(Y_{t-j} - \boldsymbol{\beta}' z_{t-1})^2$ and $\tilde{\phi}_t \equiv (Y_t - \boldsymbol{\beta}' z_t)/\tilde{\sigma}_t$, respectively.

Assumption 1.8 (i) For some $r \in [0, 1)$, $b_j(\boldsymbol{\eta}) = O(r^j)$ for all $j \in \mathbb{N}$ and $\boldsymbol{\eta} \in \mathcal{H}$.

(ii) For some $\nu_1, \ldots, \nu_q \in [0, 1)$, $\partial b_j(\boldsymbol{\eta})/\eta_k$, $k = 1, \ldots, q$, are of order $O(\nu_k^j)$, for $j \in \mathbb{N}$ and $\boldsymbol{\eta} \in \mathcal{H}$.

The following proposition shows that the unobservable values $\{u_s; s \leq 0\}$ have no influence on the LAN form below.

Proposition 1.7 (Lee and Taniguchi (2005)) *Under Assumptions 1.4–1.8, the log-likelihood ratio* $\Lambda_n(\boldsymbol{\theta}, \boldsymbol{\theta}_n)$ *admits, under* $H(g;\boldsymbol{\theta})$, *as* $n \to \infty$, *the stochastic expansion*

$$\Lambda_n(\boldsymbol{\theta}, \boldsymbol{\theta}_n) = (\boldsymbol{\kappa}', \boldsymbol{h}')\tilde{\boldsymbol{\Delta}}_n - \frac{1}{2}\boldsymbol{\xi}'\boldsymbol{F}\boldsymbol{\xi} + o_p(1), \tag{1.62}$$

where $\tilde{\boldsymbol{\Delta}}_n = n^{-1/2}\sum_{t=1}^{n}(\tilde{\boldsymbol{\Delta}}_{1,t}', \tilde{\boldsymbol{\Delta}}_{2,t}')'$,

$$\begin{aligned}
\tilde{\boldsymbol{\Delta}}_{1,t} &= -(z_t\tilde{\sigma}_t)(g'(\tilde{\phi}_t)/g(\tilde{\phi}_t)) - (2\tilde{\sigma}_t^2)^{-1}(\partial\tilde{\sigma}_t^2/\partial\boldsymbol{\beta})\{1 + \tilde{\phi}_t(g'(\tilde{\phi}_t)/g(\tilde{\phi}_t))\},\\
\tilde{\boldsymbol{\Delta}}_{2,t} &= -(2\tilde{\sigma}_t^2)^{-1}\{1 + \tilde{\phi}_t(g'(\tilde{\phi}_t)/g(\tilde{\phi}_t))\}(\partial/\partial\boldsymbol{\eta})\,\tilde{\sigma}_t^2.
\end{aligned}$$

Here, $\tilde{\boldsymbol{\Delta}}_n \xrightarrow{d} N(\boldsymbol{0}, \boldsymbol{F})$ *under* $H(g;\boldsymbol{\theta})$.

Next we discuss the estimation of $\boldsymbol{\theta}$. In what follows distribution law of a random vector $\boldsymbol{Y}_n$ under $P_{\boldsymbol{\theta}}^{(n)}$ is denoted by $\mathcal{L}(\boldsymbol{Y}_n|P_{\boldsymbol{\theta}}^{(n)})$, and the weak convergence to Z is denoted by $\mathcal{L}(\boldsymbol{Y}_n|P_{\boldsymbol{\theta}}^{(n)}) \xrightarrow{d} Z$. Define the class $\boldsymbol{\mathcal{A}}$ of sequences of estimators $\{\boldsymbol{S}_n\}$ of $\boldsymbol{\theta}$ as

$$\boldsymbol{\mathcal{A}} = [\{\boldsymbol{S}_n\} : \mathcal{L}\{\sqrt{n}(\boldsymbol{S}_n - \boldsymbol{\theta}_n)|P_{\boldsymbol{\theta}_n}^{(n)}\} \xrightarrow{d} Z_{\boldsymbol{\theta}}, \quad \textit{a probability distribution} \ ],$$

where $Z_{\boldsymbol{\theta}}$ depends on $\{\boldsymbol{S}_n\}$ generally. Let $\mathcal{L}$ be the class of all loss functions $l : \mathbb{R}^r \to [0, \infty)$ of the form $l(\boldsymbol{x}) = \tau(|\boldsymbol{x}|)$ which satisfies $\tau(0) = 0$ and $\tau(a) \leq \tau(b)$ if $a \leq b$. Typical examples are $l(\boldsymbol{x}) = I(|\boldsymbol{x}| > a)$ and $l(\boldsymbol{x}) = |\boldsymbol{x}|^p$, $p \geq 1$, where $I(\cdot)$ is the indicator function of $(\cdot)$.

Assume that the LAN property (1.61) holds. Then, a sequence $\{\hat{\boldsymbol{\theta}}_n\}$ of estimators of $\boldsymbol{\theta}$ is said to be a sequence of asymptotically centering estimators if

$$\sqrt{n}(\hat{\boldsymbol{\theta}}_n - \boldsymbol{\theta}) - \boldsymbol{F}^{-1}\boldsymbol{\Delta}_n = o_p(1) \quad \textit{in } P_{\boldsymbol{\theta}}^{(n)}. \tag{1.63}$$

The following proposition can be proved by following the arguments in Section 83 of Strasser (1985), Jeganathan (1995), and p. 69 of Taniguchi and Kakizawa (2000).

Proposition 1.8 *Assume that the LAN property* (1.61) *for the ARCH(∞)-SM model* (1.57) *holds, and that* $\{\boldsymbol{S}_n\} \in \mathcal{A}$. *Let* $\boldsymbol{\Delta}$ *be a random vector, distributed as* $N(\boldsymbol{0}, \boldsymbol{F})$. *Then the following statements hold:*

(i) For any $l \in \mathcal{L}$ *with* $E\{l(\boldsymbol{\Delta})\} < \infty$,

$$\liminf_{n\to\infty} E[l\{\sqrt{n}(\boldsymbol{S}_n - \boldsymbol{\theta})\}|P_{\boldsymbol{\theta}}^{(n)}] \geq E\{l(\boldsymbol{F}^{-1}\boldsymbol{\Delta})\}. \tag{1.64}$$

(ii) If

$$\limsup_{n\to\infty} E[l\{\sqrt{n}(\boldsymbol{S}_n - \boldsymbol{\theta})\}|P_{\boldsymbol{\theta}}^{(n)}] \leq E\{l(\boldsymbol{F}^{-1}\boldsymbol{\Delta})\}, \tag{1.65}$$

for a nonconstant $l \in \mathcal{L}$ *with* $E\{l(\boldsymbol{\Delta})\} < \infty$, *then* $\boldsymbol{S}_n$ *is a sequence of asymptotically centering estimators.*

From Proposition 1.8, it follows that $\{\hat{\boldsymbol{\theta}}_n\} \in \mathcal{A}$ is asymptotically efficient if it is asymptotically centering. Let us construct an asymptotically efficient estimator. For any sequence of estimators $\tilde{\boldsymbol{\theta}}_n$, the discretized estimator $\bar{\boldsymbol{\theta}}_n$ of $\tilde{\boldsymbol{\theta}}_n$ is defined by the nearest vertex of $\{\boldsymbol{\theta}; \boldsymbol{\theta} = n^{-1/2}(i_1, \ldots, i_r)', \ i_j \ integers\}$. First, we assume that the innovation density $g(\cdot)$ is known. Denote the Fisher information matrix $\boldsymbol{F}$ and the central sequence $\tilde{\boldsymbol{\Delta}}_n$ by $\boldsymbol{F}(\boldsymbol{\theta}, g)$ and $\tilde{\boldsymbol{\Delta}}_n(\boldsymbol{\theta}, g)$, respectively. Let

$$\hat{\boldsymbol{\theta}}_n = \bar{\boldsymbol{\theta}}_n + n^{-1/2}\boldsymbol{F}(\bar{\boldsymbol{\theta}}_n, g)^{-1}\tilde{\boldsymbol{\Delta}}_n(\bar{\boldsymbol{\theta}}_n, g)$$

where $\bar{\boldsymbol{\theta}}_n$ is a discrete and $\sqrt{n}$ - consistent estimator of $\boldsymbol{\theta}$ (for technical justification for use of discrete estimators, see p. 120 of Kreiss (1987). In (1.57), we can use the least squares estimator(LSE) $\hat{\boldsymbol{\beta}}_{LS}$ for $\boldsymbol{\beta}$. Then $\boldsymbol{\eta}$ is estimated by conditional LSE (see Tjøstheim (1986)),

$$\hat{\eta}(\hat{\boldsymbol{\beta}}_{LS}) = \arg\min_{\eta} \sum_{t=2}^{n} \left[(Y_t - \hat{\boldsymbol{\beta}}'_{LS} z_t)^2 - a(\eta) - \sum_{j=1}^{t-1} b_j(\eta)(Y_{t-j} - \hat{\boldsymbol{\beta}}'_{LS} z_{t-j})^2 \right]^2.$$

Then it is seen that the estimator $(\hat{\boldsymbol{\beta}}'_{LS}, \hat{\eta}'(\hat{\boldsymbol{\beta}}_{LS}))'$ becomes a candidate of $\tilde{\boldsymbol{\theta}}_n$. Similar to LeCam (1986) and Linton (1993), it can be shown that $\hat{\boldsymbol{\theta}}_n$ is asymptotically efficient. If $g(\cdot)$ is unknown, substituting an appropriate nonparametric density estimator $\hat{g}_n(\cdot)$ for $g(\cdot)$, we can propose

$$\hat{\hat{\boldsymbol{\theta}}}_n = \bar{\boldsymbol{\theta}}_n + n^{-1/2} \boldsymbol{F}(\bar{\boldsymbol{\theta}}, \hat{g}_n)^{-1} \tilde{\boldsymbol{\Delta}}_n(\bar{\boldsymbol{\theta}}_n, \hat{g}_n).$$

Next, we discuss the problem of testing. Let $\mathcal{M}(\boldsymbol{B})$ be the linear space spanned by the columns of a matrix $\boldsymbol{B}$. Consider the problem of testing the null hypothesis H, under which $\sqrt{n}(\boldsymbol{\theta} - \boldsymbol{\theta}_0) \in \mathcal{M}(\boldsymbol{B})$ for some given $r \times (r - l)$ matrix $\boldsymbol{B}$ of full rank and a given vector $\boldsymbol{\theta}_0 \in \mathbb{R}^r$. Then, similar to Sect. 8.2 of Strasser (1985) and p.78 of Taniguchi and Kakizawa (2000), it is seen that the test

$$T_n = \left\| [\boldsymbol{I}_r - \boldsymbol{F}^{1/2} \boldsymbol{B}(\boldsymbol{B}'\boldsymbol{F}\boldsymbol{B})^{-1} \boldsymbol{B}' \boldsymbol{F}^{1/2}] \boldsymbol{F}^{-1/2} \tilde{\boldsymbol{\Delta}}_n \right\|^2_{\boldsymbol{\theta}=\bar{\boldsymbol{\theta}}_n} \tag{1.66}$$

is asymptotically χ_l^2-distributed under H, and is locally asymptotic optimal. Here $\boldsymbol{I}_r$ is the $r \times r$ identity matrix.

As a very general nonlinear model, we introduced the following m-dimensional CHARN model in (1.8):

$$\boldsymbol{X}_t = \boldsymbol{F}_{\boldsymbol{\theta}}(\boldsymbol{X}_{t-1}, \ldots, \boldsymbol{X}_{t-p}) + \boldsymbol{H}_{\boldsymbol{\theta}}(\boldsymbol{X}_{t-1}, \ldots, \boldsymbol{X}_{t-q}) \cdot \boldsymbol{\epsilon}_t \tag{1.67}$$

where $\dim \boldsymbol{\theta} = r$. In what follows, without loss of generality we assume $p = q$, and set down

Assumption 1.9

(i)

$$E_{\boldsymbol{\theta}} \left\| \boldsymbol{F}_{\boldsymbol{\theta}}(\boldsymbol{X}_{t-1}, \ldots, \boldsymbol{X}_{t-p}) \right\|^2 < \infty, \quad E_{\boldsymbol{\theta}} \left\| \boldsymbol{H}_{\boldsymbol{\theta}}(\boldsymbol{X}_{t-1}, \ldots, \boldsymbol{X}_{t-p}) \right\|^2 < \infty, \quad \text{for all} \quad \boldsymbol{\theta} \in \boldsymbol{\Theta}.$$

(ii) There exists $c > 0$ such that

$$c \leq \left\| \boldsymbol{H}_{\boldsymbol{\theta}'}^{-1/2}(\boldsymbol{x}) \boldsymbol{H}_{\boldsymbol{\theta}}(\boldsymbol{x}) \boldsymbol{H}_{\boldsymbol{\theta}'}^{-1/2}(\boldsymbol{x}) \right\| < \infty,$$

for all $\boldsymbol{\theta}, \boldsymbol{\theta}' \in \boldsymbol{\Theta}$ and for all $\boldsymbol{x} \in \mathbb{R}^{mp}$.

(iii) $\boldsymbol{F_\theta}$ and $\boldsymbol{H_\theta}$ are continuously differentiable with respect to $\boldsymbol{\theta}$, and their derivatives $\partial_j \boldsymbol{F_\theta}$ and $\partial_j \boldsymbol{H_\theta}$ ($\partial_j = \partial/\partial\theta_j, j = 1, \ldots, r$) satisfy the condition that there exist square-integrable functions A_j and B_j such that

$$\|\partial_j \boldsymbol{F_\theta}\| \leq A_j \quad and \quad \|\partial_j \boldsymbol{H_\theta}\| \leq B_j, j = 1, \ldots, r, \ for \ \ all \quad \boldsymbol{\theta} \in \boldsymbol{\Theta}.$$

(iv) the innovation density $p(\cdot)$ of $\boldsymbol{\epsilon}_t$ satisfies

$$\lim_{\|\boldsymbol{u}\|\to\infty} \|\boldsymbol{u}\| p(\boldsymbol{u}) = 0, \int \boldsymbol{u}\boldsymbol{u}' p(\boldsymbol{u}) d\boldsymbol{u} = \boldsymbol{I}_m.$$

(v) the continuous derivative $\boldsymbol{D}p$ of $p(\cdot)$ exists on $\mathbb{R}^m$, and

$$\int \|p^{-1}\boldsymbol{D}p\|^4 p(\boldsymbol{u}) d\boldsymbol{u} < \infty, \quad \int \|\boldsymbol{u}\|^2 \|p^{-1}\boldsymbol{D}p\|^2 p(\boldsymbol{u}) d\boldsymbol{u} < \infty.$$

Suppose that an observed stretch $\boldsymbol{X}^{(n)} = (\boldsymbol{X}_1, \ldots, \boldsymbol{X}_n)$ from (1.67) is available. Denote by $\mathcal{P}_{n,\boldsymbol{\theta}}$ the probability distribution of $\boldsymbol{X}^{(n)}$. For two hypothetical value $\boldsymbol{\theta}$, $\boldsymbol{\theta}' \in \boldsymbol{\Theta}$, the log-likelihood ratio is

$$\begin{aligned} \Lambda_n(\boldsymbol{\theta}, \boldsymbol{\theta}') &\equiv \log \frac{d\mathcal{P}_{n,\boldsymbol{\theta}'}}{d\mathcal{P}_{n,\boldsymbol{\theta}}} \\ &= \sum_{t=p}^{n} \log \frac{p\left\{\boldsymbol{H}_{\boldsymbol{\theta}'}^{-1}\left(\boldsymbol{X}_t - \boldsymbol{F}_{\boldsymbol{\theta}'}\right)\right\} \det \boldsymbol{H}_{\boldsymbol{\theta}}}{p\left\{\boldsymbol{H}_{\boldsymbol{\theta}}^{-1}\left(\boldsymbol{X}_t - \boldsymbol{F}_{\boldsymbol{\theta}}\right)\right\} \det \boldsymbol{H}_{\boldsymbol{\theta}'}}. \end{aligned} \quad (1.68)$$

Let $H(p; \boldsymbol{\theta})$ be the hypothesis under which the concerned model is (1.67) with unknown parameter $\boldsymbol{\theta} \in \boldsymbol{\Theta}$ and the innovation density $p(\cdot)$. We introduce the sequence of contiguous alternatives by

$$\boldsymbol{\theta_n} = \boldsymbol{\theta} + \frac{1}{\sqrt{n}}\boldsymbol{h}, \quad \boldsymbol{h} \in \mathcal{S} \subset \mathbb{R}^r, \quad (1.69)$$

where $\boldsymbol{h} = (h_1, \ldots, h_r)'$ and $\mathcal{S}$ is an open set of $\mathbb{R}^r$.

Then, deriving the asymptotics of $\Lambda_n(\boldsymbol{\theta}, \boldsymbol{\theta_n})$, Kato et al. (2006) showed

Proposition 1.9 *Under Assumption* 1.9, *the family* $\left\{\mathcal{P}_{n,\boldsymbol{\theta}} : \boldsymbol{\theta} \in \boldsymbol{\Theta} \subset \mathbb{R}^r\right\}$, $n \in \mathbb{N}$, *is locally asymptotically normal.*

To estimate $\boldsymbol{\theta}$, we may use the maximum likelihood estimator given by

$$\hat{\boldsymbol{\theta}}_{ML} \equiv \arg\max_{\boldsymbol{\theta}} \Lambda\left(\underline{\boldsymbol{\theta}}, \boldsymbol{\theta}\right), \quad (1.70)$$

whose asymptotic optimality is shown in the sense of Proposition 1.8.

We often observe that financial time series data are nonstationary. When we deal with nonstationary processes, one of the difficult problems is how to set up an adequate asymptotic theory. For this Dahlhaus (1996a,b, 1997) introduced an important class of nonstationary processes and developed the statistical inference. The following definition is due to Dahlhaus (1996a,b, 1997).

Definition 1.1 A sequence of stochastic processes $X_{t,n}$ $(t = 1, \cdots, n : n \geq 1)$ is called locally stationary with transfer function A° if there exists a representation

$$X_{t,n} = \int_{-\pi}^{\pi} \exp(i\lambda t) A^\circ_{t,n}(\lambda) d\xi(\lambda) \tag{1.71}$$

where

(i) $\xi(\lambda)$ is a stochastic process on $[-\pi, \pi]$ with $\overline{\xi(\lambda)} = \xi(-\lambda)$ and

$$\text{cum}\{d\xi(\lambda_1), \ldots, d\xi(\lambda_k)\} = \eta\left(\sum_{j=1}^{k} \lambda_j\right) h_k(\lambda_1, \ldots, \lambda_{k-1}) d\lambda_1 \cdots d\lambda_{k-1},$$

where $h_1 = 0$, $h_2(\lambda) = 1/(2\pi)$, $h_k(\lambda_1, \ldots, \lambda_{k-1}) = h_k/(2\pi)^{k-1}$ for all k, and $\eta(\lambda) = \sum_{j=-\infty}^{\infty} \delta(\lambda + 2\pi j)$ is the period 2π extension of the Dirac delta function.

(ii) There exists a constant K and a 2π-periodic function $A : [0, 1] \times \mathbb{R} \to \mathbb{C}$ with $A(u, -\lambda) = \overline{A(u, \lambda)}$ and

$$\sup_{t,\lambda} \left| A^\circ_{t,n}(\lambda) - A\left(\frac{t}{n}, \lambda\right) \right| \leq K n^{-1} \tag{1.72}$$

for all n. $A(u, \lambda)$ is assumed to be continuous in u.

In what follows, we assume that $A(u, \lambda)$ depends on an unknown parameter $\boldsymbol{\theta} = (\theta_1, \ldots, \theta_r)' \in \boldsymbol{\Theta} \subset \mathbb{R}^r$, i.e., $A(u, \lambda) = A_{\boldsymbol{\theta}}(u, \lambda)$. The function $f_{\boldsymbol{\theta}}(u, \lambda) \equiv |A_{\boldsymbol{\theta}}(u, \lambda)|^2$ is called the time varying spectral density. Let $X_{1,n}, \ldots, X_{n,n}$ be realizations of a locally stationary process with $f_{\boldsymbol{\theta}}(u, \lambda)$. Writing

$$\epsilon_t = \int_{-\pi}^{\pi} \exp(it\lambda) d\xi(\lambda) \tag{1.73}$$

suppose that ϵ_t's are i.i.d.(0,1) with probability density $p(\cdot)$. Further, we assume $\{X_{t,n}\}$ has the $AR(\infty)$ representation

$$a^0_{\boldsymbol{\theta},t,n}(0)\epsilon_t = \sum_{k=0}^{\infty} b^0_{\boldsymbol{\theta},t,n}(k) X_{t-k,n}. \tag{1.74}$$

Under appropriate regularity conditions, Hirukawa and Taniguchi (2006) showed,

Proposition 1.10 *(i) The log-likelihood ratio $\Lambda_n(\boldsymbol{\theta}, \boldsymbol{\theta}_n)$ between $\boldsymbol{\theta}$ and $\boldsymbol{\theta}_n$ given in (1.69) has the LAN properties.*
(ii) For the class of regular estimators, Proposition 1.8 holds.

Since the log-likelihood ratio $\Lambda_n(\ ,\ )$ contains $\{\epsilon_s : s \leq 0\}$ which are unobservable, Hirukawa and Taniguchi (2006) introduced a quasi-maximum likelihood estimator $\hat{\boldsymbol{\theta}}_{QML}$ of $\boldsymbol{\theta}$ which is defined by maximizing the feasible one:

$$L_n(\boldsymbol{\theta}) = \prod_{t=1}^{n} \frac{1}{a^0_{\boldsymbol{\theta},t,n}(0)} p \left\{ \frac{\sum_{k=0}^{t-1} b^0_{\boldsymbol{\theta},t,n}(k) X_{t-k,n}}{a^0_{\boldsymbol{\theta},t,n}(0)} \right\}, \tag{1.75}$$

and showed that $\hat{\boldsymbol{\theta}}_{QML}$ is asymptotically efficient.

For a class of vector linear processes (1.10), assuming that the coefficient matrices satisfy long-range dependence and that $\{\boldsymbol{U}(t)\}$'s are i.i.d. with probability density $p(\cdot)$, Taniguchi and Kakizawa (2000) showed the results stated in Proposition 1.10, and proved that a feasible quasi-maximum likelihood estimator is asymptotically efficient.

1.4 Introduction to Time Series Financial Engineering

Financial engineering is the construction of various financial positions to manage financial risks. This section uses the terminology "time series financial engineering" because we will develop the arguments based on statistical inference of stochastic processes, i.e., time series analysis.

Assets are defined as contracts that give the right to receive or obligation to provide monetary cash flows (e.g., stock, bank account, bond). Stocks are called risky assets, and bank accounts and bonds are called risk-free assets.

Now the mathematical description is given. Let $(\Omega, \mathcal{F}, P)$ be a probability space, and let $\{\mathcal{F}_t\}$ be a family of sub σ-fields of $\mathcal{F}$ satisfying $\mathcal{F}_s \subset \mathcal{F}_r \ \ (s \leq r)$. Suppose that $\boldsymbol{S}_t = (S_t^0, S_t^1, \ldots, S_t^m)'$ is the price of $(m+1)$ assets at time t $(t = 0, 1, \ldots, T)$. Usually S_t^0 is taken to be a risk-free asset, and $(S_t^1, \ldots, S_t^m)'$ is taken to be a collection of m risky assets. In what follows we assume that $\{\boldsymbol{S}_t\}$ is a stochastic process on $(\Omega, \mathcal{F}, P)$, and that each $\boldsymbol{S}_t$ is $\mathcal{F}_t$-measurable.

If one invests the assets S_t^i with fraction weights $w_{i,t} (i = 0, 1, \ldots, m)$ satisfying $\sum_{i=0}^m w_{i,t}{=}1$, the fraction vector $\boldsymbol{w}_t = (w_{0,t}, w_{1,t}, \ldots, w_{m,t})'$ is called the portfolio. Here $\boldsymbol{w}_t$ is assumed to be $\mathcal{F}_{t-1}$-measurable. Then the total investment at time t is

$$v_t(\boldsymbol{w}_t) = \sum_{i=0}^{m} w_{i,t} S_t^i \tag{1.76}$$

which is called the value process. If a portfolio $\boldsymbol{w}_t$ satisfies

$$\sum_{i=0}^{m} w_{i,t-1} S_t^i = \sum_{i=0}^{m} w_{i,t} S_t^i \tag{1.77}$$

then it is said to be self-financing, which means that after the initial investment no further capital is either invested or withdrawn. A collection of assets $\boldsymbol{S}_t = (S_t^0, S_t^1, \ldots, S_t^m)'$ is said to admit an arbitrage opportunity if there exists a self-financing portfolio $\boldsymbol{w}_t$ such that

$$\begin{aligned} &\nu_0(\boldsymbol{w}_0) = 0, \qquad \nu_T(\boldsymbol{w}_T) \geq 0, \qquad (\mathrm{P}-\text{a.s.}) \\ &P\{\nu_T(\boldsymbol{w}_T) > 0\} > 0. \end{aligned} \tag{1.78}$$

If there is no self-financing portfolio for which (1.78) holds, then the collection of assets $\boldsymbol{S}_t$ is said to be arbitrage-free. "Arbitrage-free" means the impossibility of achieving a sure, strictly positive gain with a zero initial endowment.

For a self-financing portfolio $\boldsymbol{w}_t$, we have $\nu_t(\boldsymbol{w}_t) = \nu_t(\boldsymbol{w}_{t-1})$. In what follows we understand $\nu_t(\mathbf{w}_s) = \sum_{i=0}^m w_{i,s} S_t^i$. Hence, for any $l < T$, it holds that

$$\begin{aligned} \nu_T(\boldsymbol{w}_T) &= \nu_l(\boldsymbol{w}_l) + \sum_{t=l+1}^{T} [\nu_t(\boldsymbol{w}_{t-1}) - \nu_{t-1}(\boldsymbol{w}_{t-1})] \\ &= \nu_l(\boldsymbol{w}_l) + \sum_{t=l+1}^{T} \sum_{i=0}^{m} w_{i,(t-1)} (S_t^i - S_{t-1}^i). \end{aligned} \tag{1.79}$$

Let Q be the probability distribution of $\mathcal{S}_T \equiv \{\boldsymbol{S}_t : t = 0, 1, \ldots, T\}$. Suppose there exists another probability distribution Q^* of $\mathcal{S}_T$ which satisfies

(i) Q^* is equivalent to Q (i.e., $Q(A) = 0 \Leftrightarrow Q^*(A) = 0$ for any $A \in \mathcal{F}$),
(ii) $\{\boldsymbol{S}_t\}$ is a martingale with respect to Q^*, i.e.,

$$E^*\{\boldsymbol{S}_t | \mathcal{F}_{t-1}\} = \boldsymbol{S}_{t-1}, \qquad \text{a.e.}, \tag{1.80}$$

where $E^*\{\cdot\}$ is the expectation with respect to Q^*.

From (1.79) and (1.80) it is not difficult to see that, for any $l < T$,

$$E^*\{\nu_T(\boldsymbol{w}_T) | \mathcal{F}_l\} = \nu_l(\boldsymbol{w}_l) \qquad Q^* - \text{a.e.}, \tag{1.81}$$

which implies that $\nu_l(\boldsymbol{w}_l)$ is a martingale with respect to Q^*. Next we show that $\{\boldsymbol{S}_t\}$ is arbitrage-free. For this we assume that $\{\boldsymbol{S}_t\}$ is not arbitrage-free, i.e., (1.78) holds. Then,

$$\begin{aligned} &Q(\nu_0(\boldsymbol{w}_0) = 0) = 1, \\ &Q(\nu_T(\boldsymbol{w}_T) \geq 0) = 1, \\ &Q(\nu_T(\boldsymbol{w}_T) > 0) > 0. \end{aligned} \tag{1.82}$$

Since Q^* is equivalent to Q, (1.82) implies

$$\begin{aligned} Q^*(\nu_0(\boldsymbol{w}_0) = 0) &= 1, \\ Q^*(\nu_T(\boldsymbol{w}_T) \geq 0) &= 1, \\ Q^*(\nu_T(\boldsymbol{w}_T) > 0) &> 0. \end{aligned} \tag{1.83}$$

If we set $l = 0$ in (1.81), then

$$E^*\{\nu_T(\boldsymbol{w}_T)|\mathcal{F}_0\} = \nu_0(\boldsymbol{w}_0) = 0 \qquad Q^* - \text{a.e.}, \tag{1.84}$$

which contradicts (1.83). Hence, $\{\boldsymbol{S}_t\}$ must be arbitrage-free.

Proposition 1.11 (e.g., Kariya and Liu (2003)) *If a collection of assets $\boldsymbol{S}_t = (S_t^0, S_t^1, \ldots, S_t^m)'$ $(t = 0, 1, \ldots, T)$ is a martingale with respect to a probability distribution Q^* which is equivalent to the probability distribution Q of $\{\boldsymbol{S}_0, \boldsymbol{S}_1, \ldots, \boldsymbol{S}_T\}$, then the collection of assets $\boldsymbol{S}_t$ is arbitrage-free.*

The probability distribution Q^* is called the equivalent martingale measure. In Proposition 1.11, the reverse statement "if the collection of assets is arbitrage-free, there exists an equivalent martingale measure" holds.

Definition 1.2 (i) A contingent claim is a nonnegative random variable X representing a payoff at some future time T. We can regard it as a contract that an investor makes at time $t < T$ (e.g., option).

(ii) For a contingent claim X, if there exists a self-financing portfolio $\boldsymbol{w}_T$ such that

$$X = \nu_T(\boldsymbol{w}_T) \tag{1.85}$$

then $\boldsymbol{w}_T$ is called a replicating portfolio of X.

(iii) For any contingent claim, if there exists the replicating portfolio, then the market of financial assets is said to be complete.

In the case of arbitrage-free, the statement "the market is complete if and only if there exists a unique equivalent martingale measure" holds.

A derivative is a financial instrument whose value is derived from the value of some underlying instrument such as stock price, interest rate, or foreign exchange rate. Options are one example of many derivatives on the market. A call option gives one the right to buy the underlying asset by a certain date, called the maturity, for a certain price, called the strike price, while a put option gives one the right to sell the underlying asset by a maturity for a strike price. American options can be exercised at any time up to the maturity, but European options can be exercised only at their maturity. European options are easier to price than American options since one does not need to consider the possibility of early exercise. Most of the options traded on exchanges are American.

In what follows we explain these options concretely. Let S_t be the price of an underlying asset at time t. A European call with maturity T and strike price K on this asset is theoretically equivalent to a contingent claim

$$C_T = \max(S_T - K, 0). \tag{1.86}$$

If $S_T > K$ at maturity T, we will buy the asset for the strike price K and sell it immediately in the market, whence the gain is $S_T - K$. On the other hand, if $S_T \leq K$, we will not exercise the right, whence the gain is 0. But, actually, we have to pay the initial cost of purchasing the options, which is called the premium. Therfore the essential gain is $C_T -$ (premium). A European put with maturity T and strike price K is equivalent to a contingent claim

$$P_T \equiv \max(K - S_T, 0). \tag{1.87}$$

European options can only be exercised at maturity date T, but American options can be exercised at any time before or at the maturity date T. An American call option is equivalent to a contingent claim

$$C_{n^*} \equiv \max(S_{n^*} - K, 0), \tag{1.88}$$

where n^* is not determined in advance but depends on the path of the underlying process, i.e., n^* is a random variable such that the event $\{n^* = n\} \in \mathcal{F}_n = \sigma(S_1, S_2, \ldots, S_n)$, $n \leq T$. Similarly, we can define an American put option. Although there are various options, finally we just mention an Asian call option whose payoff function is given by

$$\max\left[T^{-1}\sum_{t=1}^{T} S_t - K,\ 0\right]. \tag{1.89}$$

Next we consider the problem of pricing options. We denote the present time and the maturity date of options by t and T, respectively. For a contingent claim X, assume that there exists a replicating portfolio $\boldsymbol{w}_t$ of X constructed on an asset process $\boldsymbol{S}_t$, i.e.,

$$X = v_T(\boldsymbol{w}_T). \tag{1.90}$$

Suppose that $\boldsymbol{S}_t$ is arbitrage-free. Then, converting values of future payments into their present values by risk-free interest rate r we obtain

$$E^*\{e^{-r(T-t)}X|\mathcal{F}_t\} = v_t(\boldsymbol{w}_t), \tag{1.91}$$

where $E^*\{\cdot\}$ is the expectation with respect to an equivalent martingale measure Q^*. Therefore, if X is an option, and if there exists a replicating portfolio of it, then (1.91) implies that the initial capital of the portfolio is

$$E^*\{e^{-r(T-t)}X|\mathcal{F}_t\}. \tag{1.92}$$

Hence, the reasonable price of the option X should be (1.92). Concrete valuation of (1.92) has been done for the geometric Brownian motion:

$$S_t = S_0 \exp\left[\mu t + \sigma \int_0^t dW_u\right], \qquad (t \in [0, T]), \tag{1.93}$$

where $\{W_t\}$ is the Wiener process. Dividing $[0, T]$ into N subintervals with length h, Kariya and Liu (2003) introduced a discretized version of (1.93):

$$S_n = S_0 \exp\left\{\mu nh + \sigma \sum_{k=1}^{n} u_k \sqrt{h}\right\}, \qquad (\{u_k\} \sim \text{i.i.d.} N(0, 1)), \tag{1.94}$$

where $n = 0, 1, \ldots, N$ and $Nh = T$. Then they evaluated the price of a European call option (1.86) at time $t = nh$;

$$C = \exp\{-r(T - t)\} E^* \{\max(S_N - K, 0) | \mathcal{F}_n\} \tag{1.95}$$

as follows. From (1.94), it is seen that

$$S_n = S_{n-1} \exp\left\{\mu h + \sigma \sqrt{h} u_n\right\}, \tag{1.96}$$

hence,

$$\frac{S_n}{\exp(rnh)} = \frac{S_{n-1}}{\exp\{r(n-1)h\}} \exp\left\{(\mu - r)h + \sigma \sqrt{h} u_n\right\}. \tag{1.97}$$

If the discounted process $S_n / \exp(rnh)$ becomes a martingale with respect to an equivalent martingale measure Q^*, it should hold that

$$E^*\left[\exp\{(\mu - r)h + \sigma \sqrt{h} u_n\} | \mathcal{F}_{n-1}\right] = 1 \quad \text{a.e.} \tag{1.98}$$

For this, letting the distribution of u_n under Q^* be $N(m, 1)$, we can see that the left-hand side of (1.98) is

$$\exp\{(\mu - r)h\} \times \exp\left\{m\sigma\sqrt{h} + \frac{\sigma^2}{2} h\right\}, \tag{1.99}$$

which implies that if we take

$$m = -\frac{1}{\sigma\sqrt{h}}\left\{(\mu - r)h + \frac{\sigma^2 h}{2}\right\}, \tag{1.100}$$

then (1.98) holds. Let $u_n^* = u_n - m$, then $u_n^* \sim N(0, 1)$ under Q^*. From (1.96) and (1.100) it follows that

$$S_n = S_{n-1} \exp\left\{\left(rh - \frac{\sigma^2 h}{2}\right) + \sigma\sqrt{h}u_n^*\right\}, \tag{1.101}$$

under Q^*. Recursively, we obtain

$$\begin{aligned} S_N &= S_n \exp\left\{\left(r - \frac{\sigma^2}{2}\right)(N-n)h + \sigma\sqrt{h}\sum_{j=n+1}^{N} u_j^*\right\} \\ &= S_n \exp\{A + BZ\}, \end{aligned} \tag{1.102}$$

whence

$$\begin{aligned} A &= (r - \frac{\sigma^2}{2})(N-n)h, \\ B &= \sigma\sqrt{(N-n)h}, \\ Z &= \frac{1}{\sqrt{N-n}} \sum_{j=n+1}^{N} u_j^* \sim N(0,1) \quad \text{under } Q^*. \end{aligned}$$

Then we can evaluate the call option (1.95) by (1.102), leading to the Black-Scholes formula (Black and Scholes (1973)):

$$C = S_n\, \Phi(d_t) - \exp\{-(T-t)r\}K\, \Phi(d_t - \sigma\sqrt{T-t}), \tag{1.103}$$

where $d_t = \{\log \frac{S_n}{K} + (r + \frac{\sigma^2}{2})(T-t)\}/(\sigma\sqrt{T-t}),\ T = Nh,\ t = nh$ and $\Phi(\cdot)$ is the distribution funcition of N(0, 1).

Let us return to (1.94) with $n = N$ and $h = T/N$, i.e.,

$$S_N = S_0 \exp\left\{\mu T + \sigma\sqrt{\frac{T}{N}}\sum_{k=1}^{N} u_k\right\}. \tag{1.104}$$

As we saw in Sect. 1.2, it is natural to assume that financial returns are non-Gaussian and dependent. In view of this, Tamaki and Taniguchi (2007) derived the third-order Edgeworth expansion of the distribution of $\sqrt{\frac{T}{N}}\sum_{k=1}^{N} u_k$ assuming that T is fixed and $N \nearrow \infty$, and that $\{u_k\}$ is a non-Gaussian fourth-order stationary process. Based on the Edgeworth expansion they evaluated (1.95) as in the form of

$$A_1 + N^{-1/2}A_2 + N^{-1}A_3 + o(N^{-1}), \tag{1.105}$$

without martingale adjustment. Next, imposing the martingale constraint on $S_n/\exp(rnh)$, Tamaki and Taniguchi (2007) derived an Edgeworth approximation for the arbitrage-free price (1.95) in the form:

$$C = M_1 + N^{-1/2}M_2 + N^{-1}M_3 + o(N^{-1}), \tag{1.106}$$

where M_1 equals the Black-Scholes formula. They discussed the influence of non-Gaussianity and dependence of $\{u_k\}$ on the higher order terms M_2 and M_3.

In the field of financial engineering the problem of credit rating is an important one. Usually, the credit rating has been done by use of the discriminant analysis in i.i.d. settings, although financial data are often supposed to be non-Gaussian dependent. For dependent observations the discriminant analysis has been developed by Shumway and Unger (1974), Zhang and Taniguchi (1994), and Kakizawa et al. (1998). For locally stationary processes, Sakiyama and Taniguchi (2004) addressed the problem of classfication. Because financial data are often nonstationary, in what follows, we state discrimination and classification for dependent observations in line with Sakiyama and Taniguchi (2004).

Recall that we already introduced a scalar-valued locally stationary process in (1.71). Here we generalize it to the case of vector-valued.

Definition 1.3 A sequence of vector-valued stochastic processes $\boldsymbol{X}_{t,n} = (X_{t,n}^{(1)}, \ldots, X_{t,n}^{(d)})'(t = 1, \ldots, n)$ is called locally stationary with transfer function matrix $\boldsymbol{A}_{t,n}(\lambda)=\{A_{t,n}(\lambda)_{ab}: a, b = 1, \ldots, d\}$ and mean $\boldsymbol{0}$ if there exists a representation

$$\boldsymbol{X}_{t,n} = \int_{\pi}^{\pi} \exp(i\lambda t)\boldsymbol{A}_{t,n}(\lambda)d\boldsymbol{\xi}(\lambda), \tag{1.107}$$

where the following holds:
(i) $\boldsymbol{\xi}(\lambda) = (\xi_1(\lambda), \ldots, \xi_d(\lambda))'$ is a complex-valued vecctor process on $[-\pi, \pi]$ with $\overline{\xi_a(\lambda)} = \xi_a(-\lambda)$, $E\{\xi_{a_j}(\lambda)\} = 0$ and

$$\mathrm{cum}\{d\xi_{a_1}(\lambda_1), \ldots, d\xi_{a_k}(\lambda_k)\} = \eta\left(\sum_{j=1}^{k}\lambda_j\right) g_{a_1,\ldots,a_k}(\lambda_1, \ldots, \lambda_{k-1})d\lambda_1 \cdots d\lambda_k \tag{1.108}$$

for all $a_1, \ldots, a_k \in \{1, \ldots, d\}$ and $\eta(\lambda) = \sum_{l=-\infty}^{\infty} \delta(\lambda + 2\pi l)$, ($\delta(\cdot)$ is the delta function).
(ii)There exists a constant K and a 2π-periodic matrix-valued function $\boldsymbol{A}(u, \lambda) = \{A(u, \lambda)_{a,b} : a, b = 1, \ldots, d\} : [0, 1] \times \mathbb{R} \to \mathbb{C}^{d\times d}$ with $\overline{\boldsymbol{A}(u, \lambda)} = \boldsymbol{A}(u, -\lambda)$ and

$$\sup_{t,\lambda}\left|A_{t,n}(\lambda)_{a,b} - A\left(\frac{t}{n}, \lambda\right)_{a,b}\right| \le Kn^{-1} \tag{1.109}$$

for all $a, b \in \{1, \ldots, d\}$ and $n \in \mathbb{N}$, where $\boldsymbol{A}(u, \lambda)$ is assumed to be continuous in u.

We call $\boldsymbol{f}(u, \lambda) \equiv \boldsymbol{A}(u, \lambda)\boldsymbol{A}(u, \lambda)^*$ the time-varying spectral density matrix of $\{\boldsymbol{X}_{t,n}\}$. Letting

$$d_N^{(a)}(u, \lambda) = \sum_{s=0}^{N-1} X_{[un]-N/2+s+1,n}^{(a)} \exp(-i\lambda s) \tag{1.110}$$

we introduce the periodogram matrix $\boldsymbol{I}_N(u, \lambda) = \{I_N(u, \lambda)_{a,b} : a, b = 1, \ldots, d\}$ over a segment of length N with midpoint $[un]$, where

$$I_N(u, \lambda)_{a,b} = \frac{1}{2\pi N} d_N^{(a)}(u, \lambda) d_N^{(b)}(u, -\lambda) \tag{1.111}$$

The shift from segment to segment is denoted by N. $\boldsymbol{I}_N(u_j, \lambda)$ is calculated over segments with midpoints $u_j n = t_j = N(j-1/2) \ \ (j = 1, \ldots, M)$, where $n = NM$.

For $\boldsymbol{\phi} : [0, 1] \times [-\pi, \pi] \to \mathbb{C}^{d \times d}$, define

$$J_n(\boldsymbol{\phi}) \equiv \frac{1}{M} \sum_{j=1}^{M} \int_{-\pi}^{\pi} tr\{\boldsymbol{\phi}(u_j, \lambda) \boldsymbol{I}_N(u_j, \lambda)\} d\lambda \tag{1.112}$$

and

$$J(\boldsymbol{\phi}) \equiv \int_0^1 \int_{-\pi}^{\pi} tr\{\boldsymbol{\phi}(u, \lambda) \boldsymbol{f}(u, \lambda)\} du d\lambda \tag{1.113}$$

Under appropriate regularity conditions, Dahlhaus (1997) showed that

$$(i) \qquad J_n(\boldsymbol{\phi}) \xrightarrow{P} J(\boldsymbol{\phi}), \ \ (n \to \infty), \tag{1.114}$$

$$(ii) \sqrt{n}\{J_n(\boldsymbol{\phi}) - J(\boldsymbol{\phi})\} \xrightarrow{\mathcal{L}} \mathrm{N}(0, V), (n \to \infty), \tag{1.115}$$

where V is written in terms of $\boldsymbol{\phi}$, $f(u, \lambda)$ and $g_{a_1, \ldots, a_4}(\cdot)$.

Next, we discuss the problem of classifying a vector-valued locally stationary process $\{\boldsymbol{X}_{t,n}\}$ into one of two categories described by two hypotheses:

$$\Pi_1 : \boldsymbol{f}(u, \lambda), \qquad\qquad \Pi_2 : \boldsymbol{g}(u, \lambda), \tag{1.116}$$

where $\boldsymbol{f}(u, \lambda)$ and $\boldsymbol{g}(u, \lambda)$ are $d \times d$ time-varying spectral density matrices. For this we use

$$D(\boldsymbol{f}:\boldsymbol{g}) \equiv \frac{1}{4\pi M}\sum_{j=1}^{M}\int_{-\pi}^{\pi}\left[\log\left\{\frac{|\boldsymbol{g}(u_j,\lambda)|}{|\boldsymbol{f}(u_j,\lambda)|}\right\}+tr[\boldsymbol{I}_N(u_j,\lambda)\{\boldsymbol{g}^{-1}(u_j,\lambda)-\boldsymbol{f}^{-1}(u_j,\lambda)\}]\right] \tag{1.117}$$

as a classification statistic. That is, if $D(\boldsymbol{f}:\boldsymbol{g}) > 0$ we choose category Π_1. Otherwise we choose category Π_2.

If we use $D(\boldsymbol{f}:\boldsymbol{g})$ as a classification criterion, the misclassification probabilities are $P(2|1) = P\{D(\boldsymbol{f}:\boldsymbol{g}) \le 0|\ \Pi_1\}$, $P(1|2) = P\{D(\boldsymbol{f}:\boldsymbol{g}) > 0|\ \Pi_2\}$. Then, our classification statistic is asymptotically consistent, i.e.,

Proposition 1.12 (Sakiyama and Taniguchi (2004))

$$\lim_{n\to\infty} P(2|1) = 0, \qquad \lim_{n\to\infty} P(1|2) = 0 \tag{1.118}$$

Proposition 1.12 implies that the classification criterion $D(\boldsymbol{f}:\boldsymbol{g})$ has the fundamental goodness. To evaluate more delicate goodness of $D(\boldsymbol{f}:\boldsymbol{g})$ we assume that $\boldsymbol{g}(u,\lambda)$ is contiguous to $\boldsymbol{f}(u,\lambda)$. Now we set the spectral densities as

$$\Pi_1 : \boldsymbol{f}(u,\lambda) = \boldsymbol{f}(u,\lambda), \qquad \Pi_2 : \boldsymbol{g}(u,\lambda) = \boldsymbol{f}(u,\lambda) + n^{-1/2}\boldsymbol{h}(u,\lambda), \tag{1.119}$$

where $\boldsymbol{h}(u,\lambda)$ is a $d\times d$ matrix-valued function, and assumed to be nonnegative definite on $[0,1]\times[-\pi,\pi]$.

Proposition 1.13 (Sakiyama and Taniguchi (2004)) *Suppose that $\boldsymbol{f}(u,\lambda)$ and $\boldsymbol{h}(u,\lambda)$ are continuous on $[0,1]\times[-\pi,\pi]$, and that N and n fulfill the relations $n^{1/4} \ll N \ll n^{1/2}/\log n$. Then, under (1.119),*

$$\lim_{n\to\infty} P(2|1) = \lim_{n\to\infty} P(1|2) = \Phi\left[\frac{-F/2}{\{F+H\}^{1/2}}\right], \tag{1.120}$$

where $\Phi(\cdot)$ is the distribution function of $N(0,1)$ and

$$F = \frac{1}{4\pi}\int_0^1\int_{-\pi}^{\pi} tr\{\boldsymbol{h}(u,\lambda)\boldsymbol{f}^{-1}(u,\lambda)\}^2 du\,d\lambda,$$

$$\begin{aligned}H = \frac{1}{8\pi}\int_0^1 &\sum_{b_1,b_2,b_3,b_4=1}^{d} \\ &\times\left[\int_{-\pi}^{\pi}\boldsymbol{A}(u,\lambda)^*\{\boldsymbol{f}^{-1}(u,\lambda)\boldsymbol{h}(u,\lambda)\boldsymbol{f}^{-1}(u,\lambda)\}\boldsymbol{A}(u,\lambda)d\lambda\right]_{b_2b_1} \\ &\times\left[\int_{-\pi}^{\pi}\boldsymbol{A}(u,\mu)^*\{\boldsymbol{f}^{-1}(u,\mu)\boldsymbol{h}(u,\mu)\boldsymbol{f}^{-1}(u,\mu)\}\boldsymbol{A}(u,\mu)d\mu\right]_{b_3b_4} \\ &\times g_{b_1b_2b_3b_4}(\lambda,-\lambda,-\mu)du\end{aligned} \tag{1.121}$$

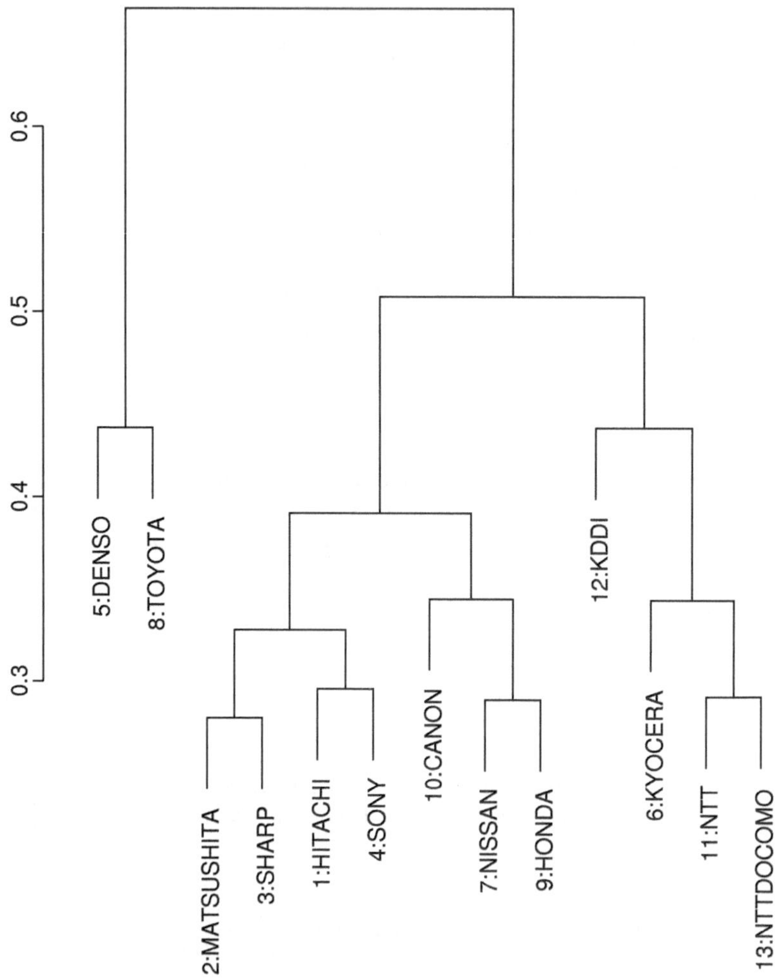

Fig. 1.4 Dendrogram for 13 financial data. Taken from Hirukawa (2006). Published with the kind permission of ©World Scientific Publishing Co. 2006. All Rights Reserved

In Proposition 1.13, the misclassification probabilities depend on H which shows an influence of non-Gaussianity of the process. Sakiyama and Taniguchi (2004) discussed non-Gaussian robustness of $D(\boldsymbol{f} : \boldsymbol{g})$ checking whether $H = 0$ or $H \neq 0$.

Suppose $\{X_{t,n}\}$ belongs to categories $\Pi_j : f(u, \lambda) = f_j(u, \lambda)$. For actual stock data, Hirukawa (2006) introduced the following classification statistic:

$$D_H(\hat{f}_j : \hat{f}_k) \equiv \frac{1}{4\pi} \int_0^1 \int_{-\pi}^{\pi} H \left\{ \frac{\hat{f}_j(u, \lambda)}{\hat{f}_k(u, \lambda)} \right\} d\lambda du, \tag{1.122}$$

where $\hat{f}_j(u, \lambda)$ is a consistent nonparametric estimator of $f_j(u, \lambda)$, and $H(\cdot)$ is a smooth function. Hirukawa (2006) executed the hierarchical clustering based on $D_H(\hat{f}_j : \hat{f}_k)$ for daily log-returns of 13 companies: 1. HITACHI, 2. MATSUSHITA, 3. SHARP, 4. SONY, 5. DENSO, 6. KYOCERA, 7. NISSAN, 8. TOYOTA, 9. HONDA, 10. CANON, 11. NTT, 12. KDDI, 13. NTTDOCOMO.

Figure 1.4 shows the dendrogram for $H(z) = z - \log z - 1$. The figure classifies the type of industry clearly, which implies that the method will be useful for the problem of credit rating.

References

Black, F., Scholes, M.: The pricing of options and corporate liabilities. J. Polit. Econom. **81**, 637–654 (1973)

Bollerslev, T.: Generalized autoregressive conditional heteroskedasticity. J. Econometrics **31**(3), 307–327 (1986)

Brillinger, D.R.: Time Series: Data Analysis and Theory, expanded edn. Holden-Day, San Francisco (2001)

Brockwell, P.J., Davis, R.A.: Time Series: Theory and Methods, 2nd edn. Springer, New York (1991)

Brown, B.M.: Martingale central limit theorems. Ann. Math. Stat. **42**, 59–66 (1971)

Chen, M., An, H.Z.: A note on the stationarity and the existence of moments of the GARCH model. Stat. Sinica **8**, 505–510 (1998)

Dahlhaus, R.: Maximum likelihhood estimation and model selection for nonstationary processes. J. Nonparametric Stat. **6**, 171–191 (1996a)

Dahlhaus, R.: On the Kullback-Leibler information divergence of locally stationary processes. Stoch. Process. Appl. **62**(1), 139–168 (1996b)

Dahlhaus, R.: Fitting time series models to nonstationary processes. Ann. Stat. **25**, 1–37 (1997)

Dunsmuir, W., Hannan, E.J.: Vector linear time series model. Adv. Appl. Probab. **8**, 339–364 (1976)

Engle, R.F.: Autoregressive conditional heteroscedasticity with estimates of the variance of United Kingdom inflation. Econometrica **50**, 987–1007 (1982)

Garel, B., Hallin, M.: Local asymptotic normality of multivariate ARMA processes with a linear trend. Ann. Inst. Stat. Math. **47**(3), 551–579 (1995)

Geweke, J.: Measurement of linear dependence and feedback between multiple time series. J. Am. Stat. Assoc. **77**, 304–313 (1982)

Giraitis, L., Surgailis, D.: A central limit theorem for quadratic forms in strongly dependent linear variables and its application to asymptotical normality of Whittle's estimate. Probab. Theory Relat. Fields **86**(1), 87–104 (1990)

Giraitis, L., Kokoszka, P., Leipus, R.: Stationary ARCH models: dependence structure and central limit theorem. Econometric Theory **16**, 3–22 (2000)

Gouriéroux, C., Jasiak, J.J.: Financial Econometrics: Problems, Models, and Methods. Princeton University Press, Princeton (2001)

Hallin, M., Ingenbleek, J.F., Puri, M.L.: Linear serial rank tests for randomness against ARMA alternatives. Ann. Stat. **13**, 1156–1181 (1985)

Hannan, E.J.: Multiple Time Series. Wiley, New York (1970)

Härdle, W., Tsybakov, A., Yang, L.: Nonparametric vector autoregression. J. Stat. Plann. Infer. **68**(2), 221–245 (1998)

Hirukawa, J.: Cluster analysis for non-Gaussian locally stationary processes. Int. J. Theor. Appl. Finance **9–1**, 113–132 (2006)

Hirukawa, J., Taniguchi, M.: LAN theorem for non-Gaussian locally stationary processes and its applications. J. Stat. Plann. Infer. **136**(3), 640–688 (2006)

Hosoya, Y., Taniguchi, M.: A central limit theorem for stationary processes and the parameter estimation of linear processes. Ann. Stat. **10**, 132–153 (1982). Correction: **21**, 1115–1117 (1993).

Jeganathan, P.: Some aspects of asymptotic theory with applications to time series models. Econometric Theory **11**(05), 818–887 (1995)

Kakizawa, Y.: Discriminant analysis for non-Gaussian vector stationary processes. J. Nonparametric Stat. **7**(2), 187–203 (1996)

Kakizawa, Y., Shumway, T.H., Taniguchi, M.: Discrimination and clustering for multivariate time series. J. Am. Stat. Assoc. **93**, 328–340 (1998)

Kariya, T., Liu, R.Y.: Asset Pricing: Discrete Time Approach. Kluwer Academic Publishers, Boston (2003)

Kato, H., Taniguchi, M., Honda, M.: Statistical analysis for multiplicatively modulated nonlinear autoregressive model and its applications to electrophysiological signal analysis in humans. IEEE Trans. Signal Process. **54**(9), 3414–3425 (2006)

Kreiss, J.P.: On adaptive estimation in stationary ARMA processes. Ann. Stat. **15**, 112–133 (1987)

Kreiss, J.P.: Local asymptotic normality for autoregression with infinite order. J. Stat. Plann. Infer. **26**(2), 185–219 (1990)

Le Cam, L.M.: Asymptotic Methods in Statistical Decision Theory. Springer, New York (1986)

Lee, S., Taniguchi, M.: Asymptotic theory for ARCH-SM models: LAN and residual empirical processes. Stat. Sinica **15**, 215–234 (2005)

Liggett Jr, W.: On the asymptotic optimality of spectral analysis for testing hypotheses about time series. Ann. Math. Stat. **42**, 1348–1358 (1971)

Ling, S., Li, W.: On fractionally integrated autoregressive moving-average time series models with conditional heteroscedasticity. J. Am. Stat. Assoc. **92**, 1184–1194 (1997)

Ling, S., McAleer, M.: On adaptive estimation in nonstationary ARMA models with GARCH errors. Ann. Stat. **31**, 642–674 (2003)

Linton, O.: Adaptive estimation in ARCH models. Econometric Theory **9**(4), 539–569 (1993)

Lu, Z., Jiang, Z.: L_1 geometric ergodicity of a multivariate nonlinear AR model with an ARCH term. Stat. Probab. Lett. **51**(2), 121–130 (2001)

Magnus, J.R., Neudecker, H.: Matrix differential calculus with applications in statistics and econometrics. Wiley, New York (1988)

Nelson, D.B.: Conditional heteroskedasticity in asset returns: a new approach. Econometrica **59**, 347–370 (1991)

Sakiyama, K., Taniguchi, M.: Discriminant analysis for locally stationry processes. J. Multivariate Anal. **90**, 282–300 (2004)

Shumway, R.H., Unger, A.N.: Linear discriminant functions for stationary time series. J. Am. Stat. Assoc. **69**, 948–956 (1974)

Strasser, H.: Mathematical Theory of Statistics. Walter de Gruyter, Berlin (1985)

Swensen, A.R.: The asymptotic distribution of the likelihood ratio for autoregressive time series with a regression trend. J. Multivariate Anal. **16**(1), 54–70 (1985)

Tamaki, K., Taniguchi, M.: Higher order asymptotic option valuation for non-Gaussian dependent return. J. Statist. Plan. Inf. Special Issue Honor of Professor Madan L. Puri. **137**, 1043–1058 (2007).

Taniguchi, M.: On estimation of the integrals of the fourth order cumulant spectral density. Biometrika **69**(1), 117–122 (1982)

Taniguchi, M., Hirukawa, J., Tamaki, K.: Optional statistical inference in financial engineering. In: Financial Mathematics Series. Chapman & Hall, New York (2008)

Taniguchi, M., Kakizawa, Y.: Asymptotic Theory of Statistical Inference for Time Series. Springer, New York (2000)

Taniguchi, M., Puri, M.L., Kondo, M.: Nonparametric approach for non-Gaussian vector stationary processes. J. Multivariate Anal. **56**(2), 259–283 (1996)

Tjøstheim, D.: Estimation in nonlinear time series models. Stoch. Process. Appl. **21**(2), 251–273 (1986)
Van der Vaart, A.W.: Asymptotic Statistics. Cambridge University Press, Cambridge (1998)
Zhang, G., Taniguchi, M.: Discriminant analysis for stationary vector time series. J. Time Ser. Anal. **15**, 117–126 (1994)

Chapter 2
Empirical Likelihood Approaches for Financial Returns

Abstract We deal with an empirical likelihood and apply it to several financial problems. Empirical likelihood is one of the nonparametric methods of statistical inference. It allows us to use likelihood methods although we do not assume that the data comes from a known family. Consequently, the empirical likelihood has both effectiveness and flexibility of the likelihood method, and reliability of the nonparametric methods. The construction of this chapter is as follows. We briefly look at the history of the empirical likelihood in Sect. 2.1 and review its method for i.i.d. data in Sect. 2.2. The frequency domain approach of empirical likelihood for multivariate non-Gaussian linear processes is discussed in Sect. 2.3. Section 2.4 gives extensions of the empirical likelihood such as Cressie-Read power-divergence statistic and generalized empirical likelihood. Section 2.5 considers application of the generalized empirical likelihood to an inference problem for multivariate stable distributions. Technical proofs of the theorems are given in Sect. 2.6.

Keywords Empirical likelihood · Non-Gaussian linear process · Frequency domain · Portfolio · Generalized empirical likelihood · Stable distribution

2.1 Introduction

Empirical likelihood is originally proposed by Owen (1988, 1990) and its overview is found in Owen (2001). Because of its advantages, many researchers investigated the empirical likelihood. For example, Qin and Lawless (1994) linked the estimating functions and equations to empirical likelihood. Newey and Smith (2004) considered generalized empirical likelihood, which is a richer class than empirical likelihood, and gave its higher order properties. For the application of empirical likelihood to the dependent data, Kitamura (1997) proposed blockwise empirical likelihood for the weakly dependent processes. Monti (1997) gave the frequency domain approach to the empirical likelihood methods for linear processes and Ogata and Taniguchi (2010)

M. Taniguchi et al., *Statistical Inference for Financial Engineering*, SpringerBriefs in Statistics, DOI: 10.1007/978-3-319-03497-3_2, extended it to the multivariate one. Ogata and Taniguchi (2009) investigated the asymptotic properties of Cressie-Read divergence statistics, which encompasses the empirical likelihood, for multivariate linear processes. A frequency domain empirical likelihood for the processes including long-range dependence is considered by Nordman and Lahiri (2006).

2.2 Empirical Likelihood Method for i.i.d. Data

Given $\boldsymbol{X}_1, \ldots, \boldsymbol{X}_n \in \mathbb{R}^m$, assumed independent with common distribution function F_0, the nonparametric likelihood of distribution function F is defined by

$$L(F) = \prod_{i=1}^{n} F(\{\boldsymbol{X}_i\}), \tag{2.1}$$

where $F(\{\boldsymbol{X}_i\})$ is the probability of getting the value $\boldsymbol{X}_i$ in a sample from F. Apparently, only the distributions which have the positive point mass probability on each observation constitute positive nonparametric likelihoods. Therefore, we restrict F to the one having the probability $p_i = F(\{\boldsymbol{X}_i\}) > 0$ on each observation $\boldsymbol{X}_i$. By a simple calculation, we find the maximizer of the nonparametric likelihood (2.1) turns to be the empirical distribution function F_n, placing probability $1/n$ on each observation. Therefore, similar to the parametric case, nonparametric likelihood ratio of F to the maximizer F_n is defined by:

$$R(F) = \frac{L(F)}{L(F_n)} = \frac{\prod_{i=1}^{n} p_i}{\prod_{i=1}^{n} 1/n} = \prod_{i=1}^{n} (np_i).$$

Suppose that we are interested in a parameter $\boldsymbol{\theta} \in \Theta \subset \mathbb{R}^p$ and that the true value $\boldsymbol{\theta}_0$ satisfies the following estimating equation

$$E[\boldsymbol{m}(\boldsymbol{X}; \boldsymbol{\theta}_0)] = \boldsymbol{0} \tag{2.2}$$

where $\boldsymbol{m}(\boldsymbol{X}; \boldsymbol{\theta}) \in \mathbb{R}^q$ is a vector-valued function, called estimating function. As examples of estimating functions, we take $\boldsymbol{m}(\boldsymbol{X}; \boldsymbol{\theta}) = \boldsymbol{X} - \boldsymbol{\theta}$ to indicate a vector mean by Eq. (2.2). For $\Pr(\boldsymbol{X} \in A)$, we take $\boldsymbol{m}(\boldsymbol{X}; \theta) = 1_{\boldsymbol{X} \in A} - \theta$. For a continuously distributed scalar X and $\theta \in \mathbb{R}$, the function $m(X; \theta) = 1_{X < \theta} - \alpha$ defines θ as the α quantile of X. Now we define the empirical likelihood ratio function for $\boldsymbol{\theta}$ by

$$\mathcal{R}(\boldsymbol{\theta}) = \max_{p} \left\{ \prod_{i=1}^{n} (np_i) \middle| \sum_{i=1}^{n} p_i \boldsymbol{m}(\boldsymbol{X}_i; \boldsymbol{\theta}) = \boldsymbol{0},\, p_i \geq 0,\, \sum_{i=1}^{n} p_i = 1 \right\}, \tag{2.3}$$

where $\boldsymbol{p} = (p_1, \ldots, p_n)$. This is the maximum of the nonparametric likelihood ratio with the restriction that the mean of the estimating function is zero under the distribution F. $\mathcal{R}(\boldsymbol{\theta})$ is calculated by the Lagrange multiplier method as follows. Write

$$G = \sum_{i=1}^{n} \log(np_i) - n\boldsymbol{\xi}' \sum_{i=1}^{n} p_i \boldsymbol{m}(\boldsymbol{X}_i; \boldsymbol{\theta}) + \gamma \left(\sum_{i=1}^{n} p_i - 1 \right),$$

where $\boldsymbol{\xi} \in \mathbb{R}^q$ and $\gamma \in \mathbb{R}$ are Lagrange multipliers. Setting $\partial G / \partial p_i = 0$ gives

$$\frac{\partial G}{\partial p_i} = \frac{1}{p_i} - n\boldsymbol{\xi}' \boldsymbol{m}(\boldsymbol{X}_i; \boldsymbol{\theta}) + \gamma = 0.$$

Therefore, the equation $\sum_{i=1}^{n} p_i (\partial G / \partial p_i) = 0$ gives $\gamma = -n$. Then, we may write

$$p_i = \frac{1}{n} \frac{1}{1 + \boldsymbol{\xi}' \boldsymbol{m}(\boldsymbol{X}_i; \boldsymbol{\theta})}$$

where the vector $\boldsymbol{\xi} = \boldsymbol{\xi}(\boldsymbol{\theta})$ satisfies q equations given by

$$\frac{1}{n} \sum_{i=1}^{n} \frac{\boldsymbol{m}(\boldsymbol{X}_i; \boldsymbol{\theta})}{1 + \boldsymbol{\xi}' \boldsymbol{m}(\boldsymbol{X}_i; \boldsymbol{\theta})} = \boldsymbol{0}. \tag{2.4}$$

Finally, we may write

$$\mathcal{R}(\boldsymbol{\theta}) = \prod_{i=1}^{n} (np_i) = \prod_{i=1}^{n} \frac{1}{1 + \boldsymbol{\xi}' \boldsymbol{m}(\boldsymbol{X}_i; \boldsymbol{\theta})}. \tag{2.5}$$

The following theorem is due to Owen (2001).

Theorem 2.1 (Owen (2001, Theorem 3.4)) *Assume that* $\mathrm{Var}(\boldsymbol{m}(\boldsymbol{X}; \boldsymbol{\theta}_0))$ *is finite and has rank* r. *Then* $-2 \log \mathcal{R}(\boldsymbol{\theta}_0) \to \chi^2_{(r)}$ *in distribution as* $n \to \infty$.

2.3 Estimation with Frequency Domain Empirical Likelihood for Stationary Processes

Here we are concerned with the m-dimensional linear process $\{\boldsymbol{X}(t)\}$ in (1.10) and consider the problem of estimating parameter $\boldsymbol{\theta} \in \Theta \subset \mathbb{R}^p$. Suppose that the information of $\boldsymbol{\theta}$ exists through a system of general estimating equations in frequency domain as follows. Let $\boldsymbol{\phi}_j(\lambda; \boldsymbol{\theta})$, $(j = 1, \ldots, q)$ be $m \times m$ matrix-valued continuous functions on $[-\pi, \pi]$ satisfying $\boldsymbol{\phi}_j(\lambda; \boldsymbol{\theta}) = \boldsymbol{\phi}_j(\lambda; \boldsymbol{\theta})^*$ and $\boldsymbol{\phi}_j(-\lambda; \boldsymbol{\theta}) = \boldsymbol{\phi}_j(\lambda; \boldsymbol{\theta})'$. We assume that each $\boldsymbol{\phi}_j(\lambda; \boldsymbol{\theta})$ satisfies the spectral moment condition

$$\int_{-\pi}^{\pi} \mathrm{tr}\{\boldsymbol{\phi}_j(\lambda; \boldsymbol{\theta}_0) \boldsymbol{f}(\lambda)\}\, d\lambda = 0 \qquad (j = 1, \ldots, q) \tag{2.6}$$

where $\boldsymbol{\theta}_0 = (\theta_{10}, \ldots, \theta_{p0})'$ is the true value of parameter and $\boldsymbol{f}(\lambda)$ is the true spectral density matrix of $\boldsymbol{X}(t)$. By taking an appropriate function for $\boldsymbol{\phi}_j(\lambda; \boldsymbol{\theta})$, the $\boldsymbol{\theta}_0$ specified by Eq. (2.6) can express various important indices for time series.

As an estimating function, we consider the following quantity

$$\boldsymbol{m}(\lambda_t; \boldsymbol{\theta}) = \Big(\mathrm{tr}\{\boldsymbol{\phi}_1(\lambda_t; \boldsymbol{\theta}) \boldsymbol{I}_n(\lambda_t)\}, \ldots, \mathrm{tr}\{\boldsymbol{\phi}_q(\lambda_t; \boldsymbol{\theta}) \boldsymbol{I}_n(\lambda_t)\}\Big)' \tag{2.7}$$

where $\lambda_t = (2\pi t)/n, (t = 1, \ldots, n)$ and $\boldsymbol{I}_n(\lambda)$ is the periodogram matrix. Following (2.3), we define the empirical likelihood ratio function for frequency domain by

$$\tilde{\mathcal{R}}(\boldsymbol{\theta}) = \max_{p} \left\{ \prod_{t=1}^{n} (np_t) \,\middle|\, \sum_{t=1}^{n} p_t \boldsymbol{m}(\lambda_t; \boldsymbol{\theta}) = \boldsymbol{0},\ p_t \geq 0, \sum_{t=1}^{n} p_t = 1 \right\}. \tag{2.8}$$

Now we impose the following assumption.

Assumption 2.1

(i) $\{\boldsymbol{X}(t)\}$ satisfies (B) in Proposition 1.2.
(ii) For the sequence $\{C_k\}$ defined by

$$C_k = \sup_{a_1 \ldots a_k} \sum_{t_1, \ldots, t_{k-1} = -\infty}^{\infty} |c^X_{a_1 \cdots a_k}(t_1, \ldots, t_{k-1})|,$$

it holds

$$\sum_{k=1}^{\infty} \frac{C_k z^k}{k!} < \infty$$

for z in a neighborhood of zero. Here, $c^X_{a_1 \cdots a_k}(t_1, \ldots, t_{k-1})$ is the k-th-order cumulant of $\boldsymbol{X}$ defined in Chap. 1.

Applying Proposition 1.2, we obtain the following theorem.

Theorem 2.2 (Ogata and Taniguchi (2010)) *Let Assumption 2.1 hold. Then*

$$-2 \log \tilde{\mathcal{R}}(\boldsymbol{\theta}_0) \to (\boldsymbol{\Sigma} \boldsymbol{N})'(\boldsymbol{\Sigma} \boldsymbol{N}),$$

where $\boldsymbol{N}$ has a q-dimensional standard normal distribution and $\boldsymbol{\Sigma} = \boldsymbol{\Sigma}_2^{-1/2} \boldsymbol{\Sigma}_1^{1/2}$. Here $\boldsymbol{\Sigma}_1$ and $\boldsymbol{\Sigma}_2$ are constant $q \times q$ matrices whose (i, j)-th components are

$$\frac{1}{\pi}\int_{-\pi}^{\pi} \mathrm{tr}\{\boldsymbol{f}(\lambda)\boldsymbol{\phi}_i(\lambda;\boldsymbol{\theta}_0)\boldsymbol{f}(\lambda)\boldsymbol{\phi}_j(\lambda;\boldsymbol{\theta}_0)\}d\lambda$$
$$+\frac{1}{2\pi}\sum_{r,t,u,v=1}^{m}\iint_{-\pi}^{\pi}\phi_{rt}^{(i)}(\lambda_1;\boldsymbol{\theta}_0)\phi_{uv}^{(j)}(\lambda_2;\boldsymbol{\theta}_0)Q_{rtuv}^{X}(-\lambda_1,\lambda_2,-\lambda_2)d\lambda_1 d\lambda_2$$

and

$$\frac{1}{2\pi}\Bigg[\int_{-\pi}^{\pi} \mathrm{tr}\{\boldsymbol{f}(\lambda)\boldsymbol{\phi}_i(\lambda;\boldsymbol{\theta}_0)\boldsymbol{f}(\lambda)\boldsymbol{\phi}_j(\lambda;\boldsymbol{\theta}_0)\}d\lambda + \int_{-\pi}^{\pi} \mathrm{tr}\{\boldsymbol{f}(\lambda)\boldsymbol{\phi}_i(\lambda;\boldsymbol{\theta}_0)\}\mathrm{tr}\{\boldsymbol{f}(\lambda)\boldsymbol{\phi}_j(\lambda;\boldsymbol{\theta}_0)\}d\lambda\Bigg],$$

respectively.

The proof of Theorem 2.2 is given in Sect. 2.6. In what follows, we give several application examples of the theorem.

Example 2.1 (Autocorrelation) Denote the autocovariance and the autocorrelation of the process $\{X_i(t)\}$ (i-th component of the process $\{\boldsymbol{X}(t)\}$) with lag h by $\gamma_i(h)$ and $\rho_i(h)$, respectively. Suppose that we are interested in the joint estimation of $\rho_i(h)$ and $\rho_j(k)$. Take

$$\boldsymbol{\phi}_1(\lambda;\boldsymbol{\theta}) = \begin{cases} \cos(h\lambda)-\theta_1 & (i,i)\text{-th component} \\ 0 & \text{otherwise} \end{cases},$$
$$\boldsymbol{\phi}_2(\lambda;\boldsymbol{\theta}) = \begin{cases} \cos(k\lambda)-\theta_2 & (j,j)\text{-th component} \\ 0 & \text{otherwise} \end{cases}.$$

Then (2.6) leads to $\theta_{10} = \gamma_i(h)/\gamma_i(0)$ and $\theta_{20} = \gamma_j(k)/\gamma_j(0)$, hence, then $\boldsymbol{\theta}_0 = (\theta_{10},\theta_{20})'$ corresponds to the desired autocorrelations $\boldsymbol{\rho} = (\rho_i(h),\rho_j(k))'$. ∎

Example 2.2 (Portfolio selection) Let $X_i(t)$ be the log-return of i-th asset $(i = 1,\ldots,m)$ at time t and suppose that the process $\{\boldsymbol{X}(t) = (X_1(t),\ldots,X_m(t))'\}$ is stationary with zero mean. Consider the portfolio $p(t) = \sum_{i=1}^{m}\theta_i X_i(t)$ where $\boldsymbol{\theta} = (\theta_1,\ldots,\theta_m)'$ is a vector of weights. The process $\{p(t)\}$ is a linear combination of the stationary process, hence $\{p(t)\}$ is also stationary and, from Herglotz's theorem, its variance is

$$\mathrm{Var}\{p(t)\} = \boldsymbol{\theta}'\mathrm{Var}\{\boldsymbol{X}(t)\}\boldsymbol{\theta} = \boldsymbol{\theta}'\left(\int_{-\pi}^{\pi}\boldsymbol{f}(\lambda)d\lambda\right)\boldsymbol{\theta}.$$

Our aim is to find the weights $\boldsymbol{\theta}_0 = (\theta_{10},\ldots,\theta_{m0})'$, that minimize the variance (the risk) of the portfolio $p(t)$. The first-order condition $(\partial/\partial\boldsymbol{\theta})_{\boldsymbol{\theta}=\boldsymbol{\theta}_0}\mathrm{Var}\{p(t)\} = \mathbf{0}$ leads to

$$\left(\int_{-\pi}^{\pi}\boldsymbol{f}(\lambda)+\boldsymbol{f}(\lambda)'d\lambda\right)\boldsymbol{\theta}_0 = \mathbf{0}. \tag{2.9}$$

Now, for $j = 1, \ldots, m$, consider to take

$$\boldsymbol{\phi}_j(\lambda; \boldsymbol{\theta}) = \begin{cases} \theta_i & (j, i)\text{-th and } (i, j)\text{-th component, } (i = 1, \ldots, m \text{ and } i \neq j) \\ 2\theta_j & (j, j)\text{-th component} \\ 0 & \text{otherwise} \end{cases}.$$

Then, (2.6) coincides with the condition (2.9), which implies that the best portfolio weights can be solved with the framework of the spectral moment condition. ∎

Example 2.3 (Whittle estimation) Consider fitting a parametric spectral density model $\boldsymbol{f}_{\boldsymbol{\theta}}(\lambda)$ to the true spectral density $\boldsymbol{f}(\lambda)$, and seeking the quasi-true value $\underline{\boldsymbol{\theta}}$ in (1.21). Assume that the specral density model has the form described in Assumption (NGR) (iii). The Kolmogorov's formula (cf. p. 162 of Hannan (1970)) says

$$\det \boldsymbol{K} = \exp\left\{ \frac{1}{2\pi} \int_{-\pi}^{\pi} \log \det\{2\pi \boldsymbol{f}_{\boldsymbol{\theta}}(\lambda)\} d\lambda \right\}.$$

This implies that, if $\boldsymbol{\theta}$ is innovation-free, the quantity $\int_{-\pi}^{\pi} \log \det\{\boldsymbol{f}_{\boldsymbol{\theta}}(\lambda)\} d\lambda$ is independent of $\boldsymbol{\theta}$ and (1.21) leads to

$$\frac{\partial}{\partial \boldsymbol{\theta}} \int_{-\pi}^{\pi} \operatorname{tr}\{\boldsymbol{f}_{\boldsymbol{\theta}}(\lambda)^{-1} \boldsymbol{f}(\lambda)\} d\lambda \bigg|_{\boldsymbol{\theta} = \underline{\boldsymbol{\theta}}} = \mathbf{0}.$$

This corresponds to (2.6) when we set

$$\boldsymbol{\phi}_j(\lambda; \boldsymbol{\theta}) = \frac{\partial \boldsymbol{f}_{\boldsymbol{\theta}}(\lambda)^{-1}}{\partial \theta_j} \quad (j = 1, \ldots, p),$$

so the quasi-true value can be expressed by the spectral moment condition. ∎

2.4 Extensions of Empirical Likelihood

Since the classical empirical likelihood method was proposed, several extensions have been considered by many researchers. Among those, we introduce Cressie-Read power-divergence (CR) statistic and generalized empirical likelihood (GEL).

As an alternative of the empirical likelihood ratio, Baggerly (1998) made use of the CR statistic, which is used at χ^2-goodness of fit test. It is defined as:

$$\frac{2}{\nu(\nu + 1)} \sum_{i=1}^{k} N_i \left\{ \left(\frac{N_i}{np_i} \right)^{\nu} - 1 \right\}, \qquad \nu \in \mathbb{R},$$

where k is a number of categories, N_i is a number of the observations which fell into the i-th category and p_i is the probability that the observation falls into the i-th category (see Read and Cressie (1988)). In the case of $\nu = -1, 0$, it is define by the continuous limits:

$$2\sum_{i=1}^{k}(np_i)\log\left(\frac{np_i}{N_i}\right) \quad \text{and} \quad 2\sum_{i=1}^{k} N_i \log\left(\frac{N_i}{np_i}\right).$$

The CR statistic contains the user-specified parameter $\nu \in \mathbb{R}$ and encompasses several commonly used test statistics, i.e., the Neyman-modified χ^2-statistic ($\nu = -2$), the Kullback-Leibler divergence ($\nu = -1$), the Freeman-Tukey statistic ($\nu = -1/2$) and Pearson's χ^2-statistic ($\nu = 1$).

In the empirical likelihood framework, we consider the distribution whose support is a set of observations. This means that $k = n$ and $N_i = 1$ for all i in the above setting. Therefore, CR statistic is reduced to

$$\mathrm{CR}_\nu(\boldsymbol{p}) = \frac{2}{\nu(\nu+1)}\sum_{i=1}^{n}\left\{(np_i)^{-\nu} - 1\right\}, \qquad \nu \in \mathbb{R}. \tag{2.10}$$

The empirical likelihood statistic corresponds to the specific case of $\nu = 0$, therefore, CR statistic can be considered as an extension of the empirical likelihood. Baggerly (1998) showed the CR statistic is asymptotically chi-square distributed for i.i.d. random vectors.

For the asymptotic distribution of CR statistic under the dependent process, we first consider the same estimating function as in (2.7). Following (2.8), we define the frequency domain CR statistic:

$$\widetilde{\mathcal{CR}}_\nu(\boldsymbol{\theta}) = \min_{\boldsymbol{p}}\left\{\mathrm{CR}_\nu(\boldsymbol{p}) \,\middle|\, \sum_{t=1}^{n} p_t \boldsymbol{m}(\lambda_t;\boldsymbol{\theta}) = \boldsymbol{0},\ p_t \geq 0, \sum_{t=1}^{n} p_t = 1\right\}. \tag{2.11}$$

Then we obtain the following theorem.

Theorem 2.3 (Ogata and Taniguchi (2009)) *Let Assumption 2.1 hold. Then* $\widetilde{\mathcal{CR}}_\nu(\boldsymbol{\theta}_0)$ *has the same asymptotic distribution as in Theorem 2.2 for any* $\nu \in \mathbb{R}$.

The proof of Theorem 2.3 is given in Sect. 2.6. In what follows, we give several application examples, in which we assume that $\boldsymbol{X}(t)$ is scalar, i.e., $\boldsymbol{X}(t) = X(t)$, for simplicity.

Example 2.4 (Prediction) The h-step prediction problem as in Hannan (1970, Chapter III Section 2) is considered. We predict $X(t)$ using a linear combination of $X(t-j)$, $j \geq h$, that is, we use $\tilde{X}(t) = \sum_{j\geq h} a(j;\boldsymbol{\theta})X(t-j)$ as a predictor where $a(j;\boldsymbol{\theta})$s are constants. We measure the error of the predictor by $E\big[|X(t) - \tilde{X}(t)|^2\big]$ and find the best linear predictor which minimizes this error. The spectral representations of $X(t)$ and $\tilde{X}(t)$ are

$$X(t) = \int_{-\pi}^{\pi} \exp(-it\lambda)\, z(d\lambda), \quad \tilde{X}(t) = \int_{-\pi}^{\pi} \exp(-it\lambda) \left\{ \sum_{j \geq h} a(j; \boldsymbol{\theta}) \exp(ij\lambda) \right\} z(d\lambda)$$

where $E\big[|z(d\lambda)|^2\big] = f(\lambda)\, d\lambda$, $E\big[z(d\lambda_1)\, \overline{z(d\lambda_2)}\big] = 0$, $\lambda_1 \neq \lambda_2$. Therefore, the prediction error is written by

$$\int_{-\pi}^{\pi} \left| 1 - \sum_{j \geq h} a(j; \boldsymbol{\theta}) \exp(ij\lambda) \right|^2 f(\lambda)\, d\lambda.$$

We find the minimizer of this error, $\boldsymbol{\theta}_0$, by the equation

$$\frac{\partial}{\partial \boldsymbol{\theta}} \int_{-\pi}^{\pi} \left| 1 - \sum_{j \geq h} a(j; \boldsymbol{\theta}) \exp(ij\lambda) \right|^2 f(\lambda)\, d\lambda \Bigg|_{\boldsymbol{\theta} = \boldsymbol{\theta}_0} = \mathbf{0}.$$

This is exactly our problem when we take

$$\phi_j(\lambda; \boldsymbol{\theta}) = \frac{\partial}{\partial \theta_j} \left| 1 - \sum_{j \geq h} a(j; \boldsymbol{\theta}) \exp(ij\lambda) \right|^2$$

in (2.6).

Example 2.5 (Interpolation) Assume that the entire time series has been observed except for the time point $t = 0$. We would like to estimate $X(0)$ by a linear combination of the observed stochastic variables, that is, $\tilde{X}(0) = \sum_{j \neq 0} a(j; \boldsymbol{\theta}) X(j)$. Similar to the prediction problem, the error of interpolation becomes

$$\int_{-\pi}^{\pi} \left| 1 - \sum_{j \neq 0} a(j; \boldsymbol{\theta}) \exp(ij\lambda) \right|^2 f(\lambda)\, d\lambda.$$

We find the minimizer of this error, $\boldsymbol{\theta}_0$, by the equation

$$\frac{\partial}{\partial \boldsymbol{\theta}} \int_{-\pi}^{\pi} \left| 1 - \sum_{j \neq 0} a(j; \boldsymbol{\theta}) \exp(ij\lambda) \right|^2 f(\lambda)\, d\lambda. \Bigg|_{\boldsymbol{\theta} = \boldsymbol{\theta}_0} = \mathbf{0}.$$

This is exactly our problem when we take

$$\phi_j(\lambda; \boldsymbol{\theta}) = \frac{\partial}{\partial \theta_j} \left| 1 - \sum_{j \neq 0} a(j; \boldsymbol{\theta}) \exp(ij\lambda) \right|^2$$

in (2.6).

Example 2.6 (Smoothing) Consider smoothing the trajectory of $X(t)$ by a linear combination of adjacent observations, $\sum_{j=-N}^{N} \theta_j X(t + j)$. Then, similar to the previous problem, the error of this smoothing is expressed as:

$$\int_{-\pi}^{\pi} \left|1 - \sum_{j=-N}^{N} \theta_j \exp(ij\lambda)\right|^2 f(\lambda)\, d\lambda.$$

Set $\boldsymbol{\theta} = (\theta_{-N}, \ldots, \theta_N)'$ and we find the minimizer of this error, $\boldsymbol{\theta}_0$, by the equation

$$\frac{\partial}{\partial \boldsymbol{\theta}} \int_{-\pi}^{\pi} \left|1 - \sum_{j=-N}^{N} \theta_j \exp(ij\lambda)\right|^2 f(\lambda)\, d\lambda \bigg|_{\boldsymbol{\theta}=\boldsymbol{\theta}_0} = \mathbf{0}.$$

This is exactly our problem when we take

$$\phi_j(\lambda; \boldsymbol{\theta}) = \frac{\partial}{\partial \theta_j} \left|1 - \sum_{j=-N}^{N} \theta_j \exp(ij\lambda)\right|^2$$

in (2.6).

Another extension of empirical likelihood is GEL in Newey and Smith (2004). GEL can be also considered as an alternative of generalized methods of moments (GMM) and it is known that its asymptotic bias does not grow with the number of moment restrictions, while the bias of GMM often does. Following the i.i.d. GEL described in Newey and Smith (2004), Ogata (2012) investigated the asymptotic properties of frequency domain GEL.

To describe GEL, let $\rho(y)$ be a function of a scalar y that is concave on its domain, an open interval $\mathcal{Y}$ containing zero. Let $\hat{\Xi}_n(\boldsymbol{\theta}) = \{\boldsymbol{\xi} : \boldsymbol{\xi}'\boldsymbol{m}(\lambda_t; \boldsymbol{\theta}) \in \mathcal{Y}, t = 1, \ldots, n\}$. The estimator is the solution to a saddle point problem

$$\hat{\boldsymbol{\theta}}_{\mathrm{GEL}} = \arg\min_{\boldsymbol{\theta} \in \Theta} \sup_{\boldsymbol{\xi} \in \hat{\Xi}_n(\boldsymbol{\theta})} \sum_{t=1}^{n} \rho\big(\boldsymbol{\xi}'\boldsymbol{m}(\lambda_t; \boldsymbol{\theta})\big).$$

The empirical likelihood (EL) estimator of Qin and Lawless (1994), the exponential tilting (ET) estimator of Kitamura and Stutzer (1997) and the continuous updating estimator (CUE) of Hansen et al. (1996) are special cases with $\rho(y) = \log(1 - y)$, $\rho(y) = -e^y$ and $\rho(y) = -(1 + y)^2/2$, respectively. Let $\boldsymbol{\Omega} = E[\boldsymbol{m}(\lambda_t; \boldsymbol{\theta}_0)\boldsymbol{m}(\lambda_t; \theta_0)']$. The following assumptions and theorems are frequency domain version of those in Newey and Smith (2004), and are found in Ogata (2012).

Assumption 2.2

(i) $\boldsymbol{\theta}_0 \in \Theta$ is the unique solution to (2.6).
(ii) Θ is compact.
(iii) $\boldsymbol{m}(\lambda_t; \boldsymbol{\theta})$ is continuous at each $\boldsymbol{\theta} \in \Theta$ with probability one.

Table 2.1 Market

Index	Market
S&P 500	NYSE
Bovespa	São Paulo stock exchange
CAC40	Bourse de Paris
AEX	Amsterdam stock exchange
ATX	Wiener Börse
HKHSI	Hong Kong exchanges and clearing
Nikkei	Tokyo stock exchange

Taken from Ogata (2012). Published with the kind permission of © Hiroaki Ogata 2012. Published under the Creative Commons Attribution License

(iv) $E[\sup_{\boldsymbol{\theta}\in\Theta} ||\boldsymbol{m}(\boldsymbol{\lambda}_t;\boldsymbol{\theta})||^{\alpha}] < \infty$ for some $\alpha > 2$.
(v) $\boldsymbol{\Omega}$ is nonsingular.
(vi) $\rho(y)$ is twice continuously differentiable in a neighborhood of zero.

Theorem 2.4 *Let Assumption 2.2 hold. Then* $\hat{\boldsymbol{\theta}}_{\text{GEL}} \xrightarrow{p} \boldsymbol{\theta}_0$.

Furthermore, let $\boldsymbol{G} = E[\partial \boldsymbol{m}(\boldsymbol{\lambda}_t;\boldsymbol{\theta}_0)/\partial\boldsymbol{\theta}]$.

Assumption 2.3

(i) $\boldsymbol{\theta}_0 \in \text{int}(\Theta)$.
(ii) $\boldsymbol{m}(\lambda_t;\boldsymbol{\theta})$ is continuously differentiable in a neighborhood $\mathcal{N}$ of $\boldsymbol{\theta}_0$ and

$$E\left[\sup_{\boldsymbol{\theta}\in\mathcal{N}}\left\|\frac{\partial \boldsymbol{m}(\lambda_t;\boldsymbol{\theta})}{\partial\boldsymbol{\theta}'}\right\|\right] < \infty.$$

(iii) $\text{rank}(\boldsymbol{G}) = p$.

Theorem 2.5 *Let Assumptions 2.2 and 2.3 hold. Then*

$$\sqrt{n}(\hat{\boldsymbol{\theta}}_{\text{GEL}} - \boldsymbol{\theta}_0) \xrightarrow{d} N(\boldsymbol{0}, (\boldsymbol{G}'\boldsymbol{\Omega}^{-1}\boldsymbol{G})^{-1}).$$

Now we are applying the frequency domain GEL estimation method to the portfolio selection problem introduced in Example 2.2. The sample consists of 7 weekly market indices (S&P 500, Bovespa, CAC 40, AEX, ATX, HKHSI, and Nikkei) having 800 observations each: the initial date is April 30, 1993, and the ending date is August 22, 2008. The data are from Ogata (2012). Refer to Table 2.1 for the market of each index.

The log-return of the i-th ($i = 1, \ldots, 7$) index at time t is denoted by $x_i(t)$ and each process is assumed to be stationary. We estimate the best portfolio weights $\boldsymbol{\theta}_0 = (\theta_1, \ldots, \theta_7)'$ described in Example 2.2 by using three types of frequency domain GEL estimators (EL, ET, and CUE). The results are shown in Table 2.2, and explain that Bovespa and ATX contribute large part in the optimal portfolio.

Table 2.2 Estimated portfolio weights

	EL	ET	CUE
S&P 500	0.0759	0.0617	0.0001
Bovespa	0.6648	0.6487	0.6827
CAC40	0.0000	0.0000	0.0000
AEX	0.0000	0.0000	0.0000
ATX	0.2593	0.2558	0.3168
HKHSI	0.0000	0.0000	0.0000
Nikkei	0.0000	0.0338	0.0004

Taken from Ogata (2012). Published with the kind permission of © Hiroaki Ogata 2012. Published under the creative commons attribution license

2.5 GEL for Multivariate Stable Distributions

It is well known that financial data often show asymmetry and heavy-tails properties which cannot be captured by simple distributions such as normal distribution. One of the families of distributions which can be accommodated to such asymmetry and heavy-tails data is the stable family. Since Mandelbrot (1963) and Fama (1965) proposed the use of stable distributions to analyze financial data, these distributions have received considerable attention; see, for example, Embrechts et al. (1997), Belkacem et al. (2000), and Rachev and Han (2000). Moreover, the stable distributions arise from a generalization of the central limit theorem in which the assumption of finite variance is relaxed, and consequently, the stable distributions are closed under summation. This property is also a strong motivation to fit stable distributions to financial data, since low frequency financial returns can be regarded as the sum of high frequency data.

In order to deal with several financial assets simultaneously, we treat multivariate stable distributions. In fact, we have multivariate stable distributions that are included in an elliptical family. Elliptical stable distributions are easier to handle than nonelliptical stable distributions but we do not consider them in this section. The reason is in Fig. 2.1, that displays density plots of bivariate normal, elliptical stable, and nonelliptical stable distributions along with contour plots of the densities. The graphs for the nonelliptical stable distribution show skewness that cannot be handled by either normal or elliptical stable distributions. Therefore, we especially consider multivariate nonelliptical stable distributions and apply the i.i.d. GEL method to estimate their parameters, following Ogata (2013).

We briefly review the multivariate stable distributions. A random vector $\boldsymbol{X} = (X_1, \ldots, X_d)' \in \mathbb{R}^d$ is said to be stable if its characteristic function is

$$\phi(\boldsymbol{\nu}) = E\big[\exp\{i\langle \boldsymbol{X}, \boldsymbol{\nu}\rangle\}\big] = \exp\{-I(\boldsymbol{\nu})\},$$

where the exponent function is

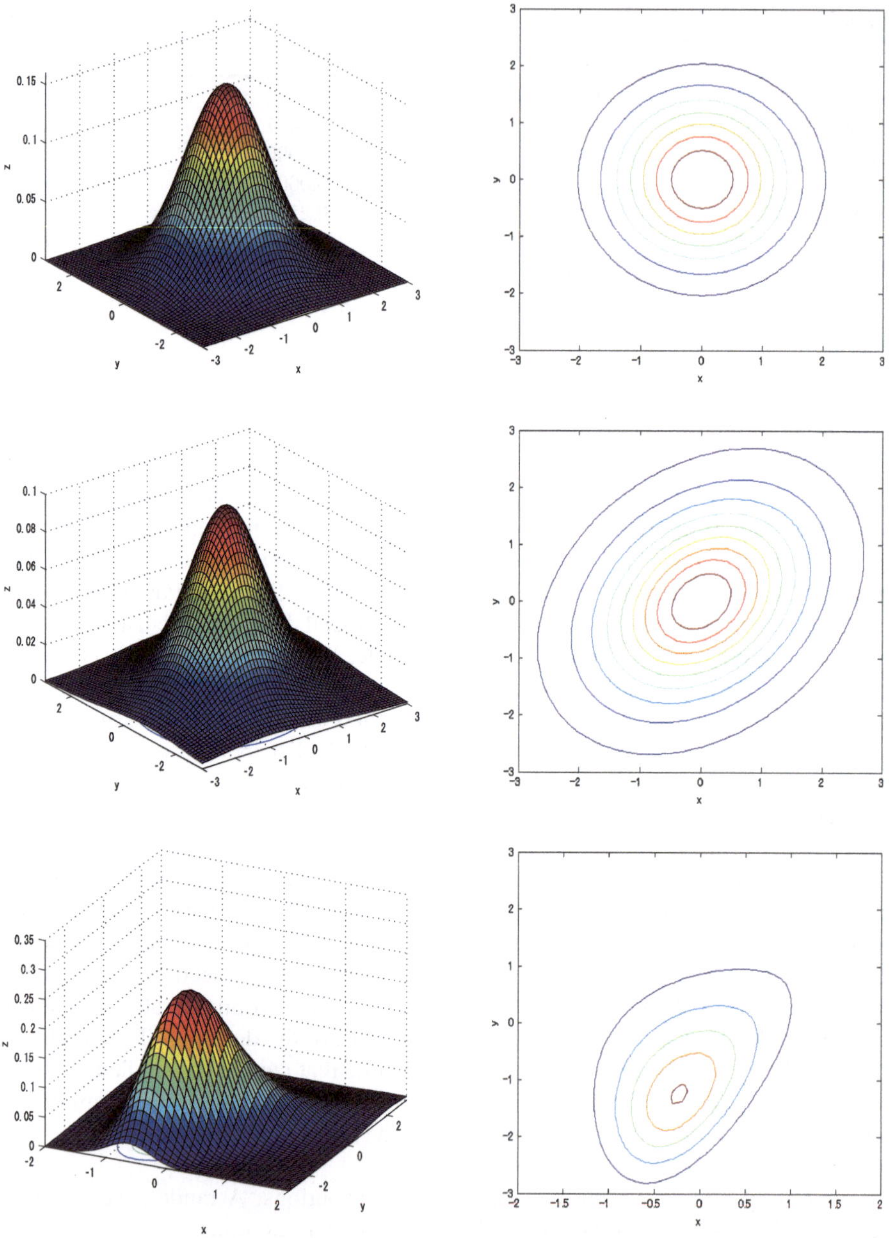

Fig. 2.1 The *left three panels* show bivariate density plots and *right three panels* show their contours. From *top* to *bottom*, the figures are for normal, elliptical stable, and nonelliptical stable distributions, respectively. Taken from Ogata (2013). Published with the kind permission of

$$I(\boldsymbol{v}) = \int_{S_d} \psi(\langle \boldsymbol{s}, \boldsymbol{v} \rangle) \Gamma(d\boldsymbol{s}) + i \langle \boldsymbol{v}, \boldsymbol{\mu} \rangle.$$

Here $S_d = \{\boldsymbol{s} : \| \boldsymbol{s} \| = 1\}$ is the unit sphere in $\mathbb{R}^d$, the symbol $\langle \cdot, \cdot \rangle$ denotes inner product, Γ is a finite spectral measure on S_d, $\boldsymbol{\mu} = (\mu_1, \ldots, \mu_d) \in \mathbb{R}^d$ is the location vector, and

$$\psi(u) = \begin{cases} |u|^\alpha \left(1 - i \operatorname{sign}(u) \tan \dfrac{\pi\alpha}{2}\right) & (\alpha \neq 1) \\ |u| \left(1 + i \dfrac{2}{\pi} \operatorname{sign}(u) \ln |u|\right) & (\alpha = 1), \end{cases}$$

where $\alpha \in (0, 2]$ is the characteristic exponent, which controls tail thickness. The d-dimensional stable distribution is denoted by $S^d(\alpha, \Gamma, \boldsymbol{\mu})$.

Our purpose is to estimate the spectral measure Γ, the location vector $\boldsymbol{\mu}$, and the characteristic exponent α, based on an i.i.d. sample $\boldsymbol{X}_1, \ldots, \boldsymbol{X}_n$ of d-dimensional random vectors drawn from this distribution. However, difficulties arise even in the univariate case, $d = 1$. Foremost is the complexity of the density function. Except for a few cases, no simple explicit form of the density exists, which is an obstacle to implementing the usual MLE method. Moreover, the stable distributions do not necessarily have second or even first moments, which impede using the ordinary method of moments.

One remedy for this difficulty is to use the empirical characteristic function

$$\hat{\phi}_n(\boldsymbol{v}) = \frac{1}{n} \sum_{j=1}^{n} \exp\{i \langle \boldsymbol{X}_j, \boldsymbol{v} \rangle\}.$$

Following Nolan et al. (2001), we first consider a discrete approximation of the spectral measure:

$$\Gamma^* = \sum_{\ell=1}^{L} \gamma_\ell \delta_{\boldsymbol{s}_\ell}, \tag{2.12}$$

where $\gamma_\ell = \Gamma(A_\ell)$, $\ell = 1, \ldots, L$, are weights and $\delta_{\boldsymbol{s}_\ell}$ is a point mass at $\boldsymbol{s}_\ell \in S_d$. Here, A_ℓ, $\ell = 1, \ldots, L$, are patches that partition the sphere S_d, where A_ℓ has some "center" $\boldsymbol{s}_\ell$. The estimation of the parameters is done via an estimating function based on theoretical and empirical characteristic functions. Many authors have investigated the use of empirical characteristic functions for estimation; for a summary and extensive references, see Yu (2004). In the context of multivariate stable distributions, suppose that $\{\boldsymbol{X}_j\}_{j=1}^n$ is an i.i.d. sequence from $S^d(\alpha, \Gamma^*, \boldsymbol{\mu})$, where Γ^* is a discrete spectral measure defined in (2.12). Denote the parameters by $\boldsymbol{\theta} = (\alpha, \boldsymbol{\gamma}', \boldsymbol{\mu}')' \in (0, 2] \times \mathbb{R}_+^L \times \mathbb{R}^d$, and define the estimating function as

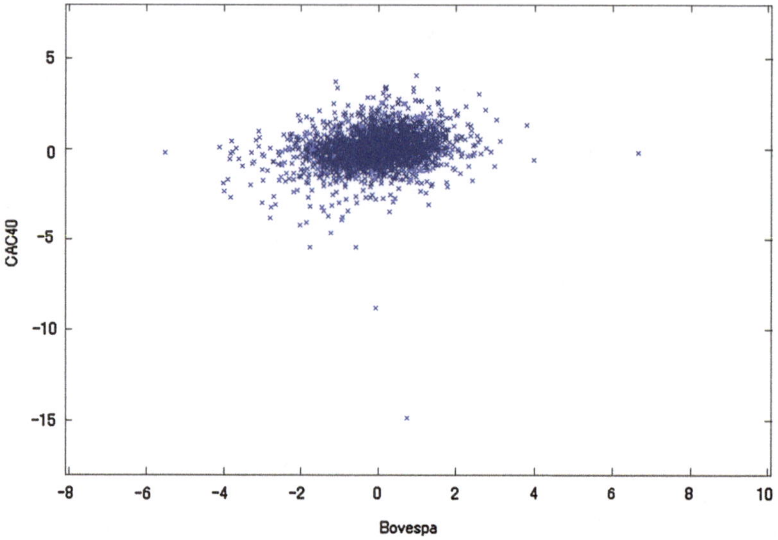

Fig. 2.2 Scatter plot of the pair of Bovespa and CAC40. Taken from Ogata (2013). Published with the kind permission of © Elsevier 2013. All Rights Reserved

$$h(\boldsymbol{\nu}, \boldsymbol{X}_j, \boldsymbol{\theta}) = \exp\{i\langle \boldsymbol{\nu}, \boldsymbol{X}_j\rangle\} - \phi_{\boldsymbol{\theta}}(\boldsymbol{\nu}),$$

where $\phi_{\boldsymbol{\theta}}$ is the theoretical characteristic function for model parameter $\boldsymbol{\theta}$. Denoting true parameter value by $\boldsymbol{\theta}_0$, we have $E[h(\boldsymbol{\nu}, \boldsymbol{X}_j, \boldsymbol{\theta}_0)] = 0$ for any frequency $\boldsymbol{\nu} \in \mathbb{R}^d$. After choosing some frequencies $\boldsymbol{\nu}_1, \ldots, \boldsymbol{\nu}_K \in \mathbb{R}^d$, we redefine the estimating function as

$$\begin{aligned} \boldsymbol{g}(\boldsymbol{X}_j, \boldsymbol{\theta}) = (&\mathfrak{Re}[h(\boldsymbol{\nu}_1, \boldsymbol{X}_j, \boldsymbol{\theta})], \ldots, \mathfrak{Re}[h(\boldsymbol{\nu}_K, \boldsymbol{X}_j, \boldsymbol{\theta})], \\ &\mathfrak{Im}[h(\boldsymbol{\nu}_1, \boldsymbol{X}_j, \boldsymbol{\theta})], \ldots, \mathfrak{Im}[h(\boldsymbol{\nu}_K, \boldsymbol{X}_j, \boldsymbol{\theta})])', \end{aligned} \tag{2.13}$$

where $\mathfrak{Re}[\cdot]$ and $\mathfrak{Im}[\cdot]$ are the real and imaginary parts of a complex number. Obviously, we still have $E[\boldsymbol{g}(\boldsymbol{X}_j, \boldsymbol{\theta}_0)] = \boldsymbol{0}$. Using this estimating function, we estimate the parameters by the GEL method described in the previous section.

Now we are applying the GEL estimation method for multivariate stable distributions to real data. The sample consists of two log returns of daily market indexes, Bovespa (São Paulo Stock Exchange) and CAC40 (Bourse de Paris), having 2536 observations in each.[1] The initial date is January 4, 2000, and the ending date is September 22, 2009. Figure 2.2 shows a scatter plot of their returns. In this plot, several points are located far from the origin, and the data are skewed heavily downward. Thus, multivariate stable distributions are more appropriate than multivariate normal or elliptical distributions.

[1] This data is a part of the data set which was used in Dominicy and Veredas (2013).

Table 2.3 Estimated parameters and VaRs : σ_{BB} and σ_{CC} are the variances for Bovespa and CAC40, and σ_{BC} is the covariance

Model	Stable				Normal	
	α	1.757				
	γ_1	0.409	γ_5	0.598		
Estimated parameters	γ_2	0.000	γ_6	0.000	σ_{BB}	0.999
	γ_3	0.000	γ_7	0.000	σ_{BC}	0.219
	γ_4	0.226	γ_8	0.729	σ_{CC}	0.988
$\mathrm{VaR_B}(0.01)$		−5.763				−2.960
$\mathrm{VaR_C}(0.01)$		−3.778				−1.603

$\mathrm{VaR_B}(0.01)$ and $\mathrm{VaR_C}(0.01)$ stand for the 1 % VaRs for Bovespa and CAC40, respectively. Taken from Ogata (2013). Published with the kind permission of © Elsevier 2013. All Rights Reserved

We model the pair (Bovespa, CAC40) using the bivariate stable distribution $S^2(\alpha, \Gamma^*, \mathbf{0})$. Here, Γ^* is a discrete spectral measure defined in (2.12), and we take $L = 8$, $\boldsymbol{s}_\ell = (\cos\theta_\ell, \sin\theta_\ell)'$ for $\theta_\ell = (\ell - 1)\pi/4$, $\ell = 1, \ldots, 8$. For comparison, the bivariate normal distribution $N\left(\begin{pmatrix}0\\0\end{pmatrix}, \begin{pmatrix}\sigma_{BB} & \sigma_{BC}\\ \sigma_{BC} & \sigma_{CC}\end{pmatrix}\right)$ is also fitted to the data. We use ET estimators for the parameters of the stable distribution, $(\alpha, \gamma_1, \ldots, \gamma_8)'$, and MLEs for the parameters of the normal distribution, $(\sigma_{BB}, \sigma_{BC}, \sigma_{CC})$. Based on the estimated parameters for both models, we calculate the 1 % VaRs for the Bovespa and the CAC40. Table 2.3 shows the estimation results and the VaRs.

Both VaRs from the fitted stable model are smaller than those from the fitted normal model, which means fitting the stable distribution evaluates the risk to be higher than that from fitting the normal distribution. The estimated marginal density functions for Bovespa and CAC40 are displayed in Fig. 2.3.

Last, we make a comment about the data. Perhaps, heavy tails may be caused by GARCH effects rather than by unconditional stable distributions. To remove the effect of the dynamic conditional volatility, Dominicy and Veredas (2013) adjusted the original data by using the AR(2)-GARCH(1,1) model. The analysis here is also based on this adjusted data.

2.6 Appendix

In this section, we provide the proofs of Theorems 2.2 and 2.3. For the simplicity of expression, we introduce

$$\boldsymbol{P}_n = \frac{1}{\sqrt{n}} \sum_{t=1}^{n} \boldsymbol{m}(\lambda_t; \boldsymbol{\theta}_0),$$

$$\boldsymbol{S}_n = \frac{1}{n} \sum_{t=1}^{n} \boldsymbol{m}(\lambda_t; \boldsymbol{\theta}_0)\boldsymbol{m}(\lambda_t; \boldsymbol{\theta}_0)',$$

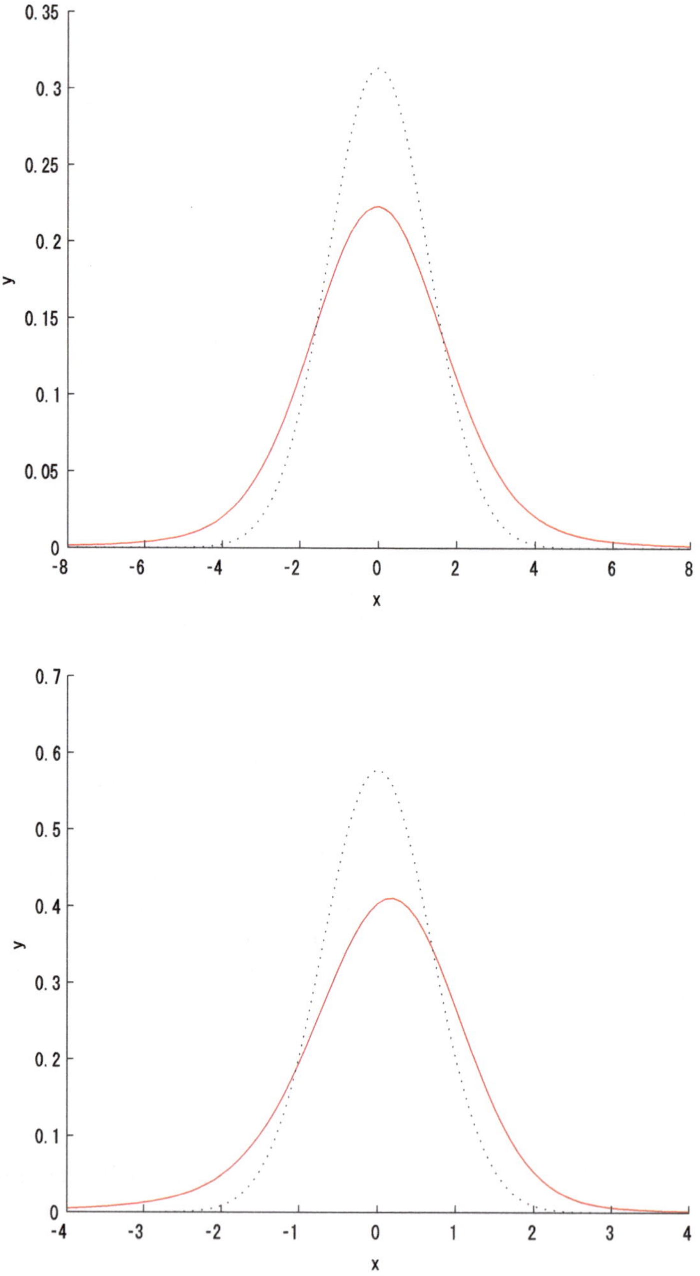

Fig. 2.3 The estimated probability density functions are displayed. The *upper figure* is for Bovespa while *lower* is for CAC40. The *real line* is for stable fitting and *dotted line* is for normal fitting. Taken from Ogata (2013). Published with the kind permission of © Elsevier 2013. All Rights Reserved

$$Z_n = \max_{1 \le t \le n} \|\boldsymbol{m}(\lambda_t; \boldsymbol{\theta}_0)\|.$$

Before giving the proofs of the theorems, we prepare Lemmas 2.1, 2.2 and 2.3 whose proofs are omitted. Lemma 2.1 is proved by use of Theorem 5.10.2 and Lemma P5.1 of Brillinger (2001), and Proposition 1.2 in Chap. 1. Lemmas 2.2 and 2.3 are proved by Theorems 7.2.2 and 4.5.1 of Brillinger (2001), respectively.

Lemma 2.1 *Let Assumption 2.1 hold. Then*

$$\boldsymbol{P}_n \xrightarrow{d} N_q(\boldsymbol{0}, \boldsymbol{\Sigma}_1)$$

where $\boldsymbol{\Sigma}_1$ *is the same matrix as in Theorem 2.2.*

Lemma 2.2 *Let Assumption 2.1 hold. Then*

$$\boldsymbol{S}_n \xrightarrow{p} \boldsymbol{\Sigma}_2$$

where $\boldsymbol{\Sigma}_2$ *is the same matrix as in Theorem 2.2.*

Lemma 2.3 *Let Assumption 2.1 hold. Then*

$$Z_n = O(\log n).$$

2.6.1 Proof of Theorem 2.2

Following the way of leading to (2.5), we may write

$$\tilde{\mathcal{R}}(\boldsymbol{\theta}_0) = \prod_{t=1}^{n} \frac{1}{1 + \boldsymbol{\xi}'\boldsymbol{m}(\lambda_t; \boldsymbol{\theta}_0)}$$

where the vector $\boldsymbol{\xi} = \boldsymbol{\xi}(\boldsymbol{\theta}_0)$ satisfies q equations given by

$$\boldsymbol{J}(\boldsymbol{\xi}) \equiv \frac{1}{n} \sum_{t=1}^{n} \frac{\boldsymbol{m}(\lambda_t; \boldsymbol{\theta}_0)}{1 + \boldsymbol{\xi}'\boldsymbol{m}(\lambda_t; \boldsymbol{\theta}_0)} = \boldsymbol{0}. \tag{2.14}$$

Put $Y_t = \boldsymbol{\xi}'\boldsymbol{m}(\lambda_t; \boldsymbol{\theta}_0)$ and write $\boldsymbol{\xi} = \|\boldsymbol{\xi}\|\boldsymbol{u}$ where $\boldsymbol{u} \in U$ is a unit vector. By substituting $1/(1 + Y_t) = 1 - Y_t/(1 + Y_t)$ into $\boldsymbol{u}'\boldsymbol{J}(\boldsymbol{\xi}) = 0$, we have

$$
\begin{aligned}
&\boldsymbol{u}'\left\{\frac{1}{n}\sum_{t=1}^{n}\boldsymbol{m}(\lambda_t;\boldsymbol{\theta}_0)\left(1-\frac{Y_t}{1+Y_t}\right)\right\}=0\\
&\boldsymbol{u}'\left(\frac{1}{n}\sum_{t=1}^{n}\boldsymbol{m}(\lambda_t;\boldsymbol{\theta}_0)\right)=\boldsymbol{u}'\left(\frac{1}{n}\sum_{t=1}^{n}\boldsymbol{m}(\lambda_t;\boldsymbol{\theta}_0)\frac{\boldsymbol{\xi}'\boldsymbol{m}(\lambda_t;\boldsymbol{\theta}_0)}{1+Y_t}\right)\\
&\boldsymbol{u}'\left(\frac{1}{n}\sum_{t=1}^{n}\boldsymbol{m}(\lambda_t;\boldsymbol{\theta}_0)\right)=\|\boldsymbol{\xi}\|\boldsymbol{u}'\left(\frac{1}{n}\sum_{t=1}^{n}\frac{\boldsymbol{m}(\lambda_t;\boldsymbol{\theta}_0)\boldsymbol{m}(\lambda_t;\boldsymbol{\theta}_0)'}{1+Y_t}\right)\boldsymbol{u}.
\end{aligned}
$$

Then, we obtain

$$
\begin{aligned}
\|\boldsymbol{\xi}\|\boldsymbol{u}'\boldsymbol{S}_n\boldsymbol{u} &\le \|\boldsymbol{\xi}\|\boldsymbol{u}'\left(\frac{1}{n}\sum_{t=1}^{n}\frac{\boldsymbol{m}(\lambda_t;\boldsymbol{\theta}_0)\boldsymbol{m}(\lambda_t;\boldsymbol{\theta}_0)'}{1+Y_t}\right)\boldsymbol{u}\cdot(1+\max_t Y_t)\\
&\le \|\boldsymbol{\xi}\|\boldsymbol{u}'\left(\frac{1}{n}\sum_{t=1}^{n}\frac{\boldsymbol{m}(\lambda_t;\boldsymbol{\theta}_0)\boldsymbol{m}(\lambda_t;\boldsymbol{\theta}_0)'}{1+Y_t}\right)\boldsymbol{u}\cdot(1+\|\boldsymbol{\xi}\|Z_n)\\
&= n^{-1/2}\boldsymbol{u}'\boldsymbol{P}_n(1+\|\boldsymbol{\xi}\|Z_n)
\end{aligned}
$$

and so

$$
\|\boldsymbol{\xi}\|\left(\boldsymbol{u}'\boldsymbol{S}_n\boldsymbol{u}-n^{-1/2}Z_n\boldsymbol{u}'\boldsymbol{P}_n\right)\le n^{-1/2}\boldsymbol{u}'\boldsymbol{P}_n.
$$

From Lemmas 2.1–2.3, we have

$$
\|\boldsymbol{\xi}\|\{O_p(1)-O(\log n)O_p(n^{-1/2})\}\le O_p(n^{-1/2}),
$$

and hence,

$$
\|\boldsymbol{\xi}\| = O_p(n^{-1/2}). \tag{2.15}
$$

After we have established an order of $\boldsymbol{\xi}$, we have from Lemma 2.3 and (2.15) that

$$
\max_{1\le t\le n}|\boldsymbol{Y}_t| = O_p(n^{-1/2})O(\log n). \tag{2.16}
$$

Now, by (2.14),

$$
\begin{aligned}
\boldsymbol{0} &= \frac{1}{n}\sum_{t=1}^{n}\boldsymbol{m}(\lambda_t;\boldsymbol{\theta}_0)\left(1-Y_t+\frac{Y_t^2}{1+Y_t}\right) \qquad (2.17)\\
&= n^{-1/2}\boldsymbol{P}_n-\boldsymbol{S}_n\boldsymbol{\xi}+\frac{1}{n}\sum_{t=1}^{n}\boldsymbol{m}(\lambda_t;\boldsymbol{\theta}_0)\frac{Y_t^2}{1+Y_t}
\end{aligned}
$$

Noting that

$$\frac{1}{n}\sum_{t=1}^{n}\|\boldsymbol{m}(\lambda_t;\boldsymbol{\theta}_0)\|^3 \le Z_n\|\boldsymbol{S}_n\| = O(\log n),$$

the final term of (2.17) has a norm bounded by

$$\frac{1}{n}\sum_{t=1}^{n}\|\boldsymbol{m}(\lambda_t,\boldsymbol{\theta}_0)\|^3\|\boldsymbol{\xi}\|^2|1+Y_t|^{-1} = O(\log n)O_p(n^{-1})O_p(1) = O_p(n^{-1}\log n).$$

Hence we can write

$$\boldsymbol{\xi} = \frac{1}{\sqrt{n}}\boldsymbol{S}_n^{-1}\boldsymbol{P}_n + \boldsymbol{\epsilon}$$

where $\boldsymbol{\epsilon} = O_p(n^{-1}\log n)$. By (2.16), we can write

$$\log(1+Y_t) = Y_t - \frac{1}{2}Y_t^2 + \eta_t$$

where, for some finite constant $C > 0$,

$$\Pr(|\eta_t| \le C|Y_t|^3, 1 \le t \le n) \to 1 \quad \text{as} \quad n \to \infty.$$

Now we may write

$$\begin{aligned}-2\log\tilde{\mathcal{R}}(\boldsymbol{\theta}_0) &= -2\sum_{t=1}^{n}\log(nw_t) = -2\sum_{t=1}^{n}\log(1+Y_t) = 2\sum_{t=1}^{n}Y_t - \sum_{t=1}^{n}Y_t^2 + 2\sum_{t=1}^{n}\eta_t \\ &= \boldsymbol{P}_n'\boldsymbol{S}_n^{-1}\boldsymbol{P}_n - n\boldsymbol{\epsilon}'\boldsymbol{S}_n\boldsymbol{\epsilon} + 2\sum_{t=1}^{n}\eta_t.\end{aligned}$$

Here it is seen that

$$n\boldsymbol{\epsilon}'\boldsymbol{S}_n\boldsymbol{\epsilon} = nO_p(n^{-1}\log n)O_p(1)O_p(n^{-1}\log n) = O_p(n^{-1}(\log n)^2),$$

$$2\sum_{t=1}^{n}\eta_t \le C\|\boldsymbol{\xi}\|^3\sum_{t=1}^{n}\|\boldsymbol{m}(\lambda_t,\boldsymbol{\theta}_0)\|^3 = O_p(n^{-3/2})O(n\log n) = O_p(n^{-1/2}\log n),$$

and finally, from Lemmas 2.1 and 2.2, we can show that

$$\boldsymbol{P}_n'\boldsymbol{S}_n^{-1}\boldsymbol{P}_n \xrightarrow{d} \left(\boldsymbol{\Sigma}_2^{-1/2}\boldsymbol{\Sigma}_1^{1/2}\boldsymbol{\Sigma}_1^{-1/2}\boldsymbol{P}_n\right)'\left(\boldsymbol{\Sigma}_2^{-1/2}\boldsymbol{\Sigma}_1^{1/2}\boldsymbol{\Sigma}_1^{-1/2}\boldsymbol{P}_n\right) \stackrel{d}{=} (\boldsymbol{\Sigma}\boldsymbol{N})'(\boldsymbol{\Sigma}\boldsymbol{N}).$$

□

2.6.2 Proof of Theorem 2.3

Let $\nu \neq 0, -1$. To find the minimizing weights $\boldsymbol{p} = (p_1, \ldots, p_n)'$ in (2.11), we proceed by the Lagrange multiplier method. Write

$$G = \mathrm{CR}_\nu(\boldsymbol{p}) + \boldsymbol{\xi}' \sum_{t=1}^{n} p_t \boldsymbol{m}(\lambda_t; \boldsymbol{\theta}_0) + \gamma \left(\sum_{t=1}^{n} p_t - 1 \right),$$

where $\boldsymbol{\xi} \in \mathbb{R}^q$ and $\gamma \in \mathbb{R}$ are Lagrange multipliers. Since the objective function $\mathrm{CR}_\nu(\boldsymbol{p})$ is convex with respect to $\boldsymbol{p}$, the solution of the first-order condition

$$\text{(i)} \ \frac{\partial G}{\partial \boldsymbol{p}} = \boldsymbol{0} \quad \text{(ii)} \ \frac{\partial G}{\partial \boldsymbol{\xi}} = \boldsymbol{0} \quad \text{(iii)} \ \frac{\partial G}{\partial \gamma} = 0 \tag{2.18}$$

gives a global minimum point. The first condition of (2.18) gives

$$p_t = \frac{1}{n} \left\{ \frac{(\nu+1)\gamma}{2n} \right\}^{-1/(1+\nu)} \left\{ 1 + \frac{\boldsymbol{\xi}'}{\gamma} \boldsymbol{m}(\lambda_t; \boldsymbol{\theta}_0) \right\}^{-1/(1+\nu)} \quad (t = 1, \ldots, n).$$

With a simpler notation, we rewrite p_t as

$$p_t = C^* \left\{ 1 + \boldsymbol{\xi}' \boldsymbol{m}(\lambda_t; \boldsymbol{\theta}_0) \right\}^{-1/(1+\nu)} \quad (t = 1, \ldots, n) \tag{2.19}$$

where C^* and $\boldsymbol{\xi}$ are certain constant and vector, respectively. From the third condition of (2.18), the constant should be

$$C^* = \left[\sum_{t=1}^{n} \left\{ 1 + \boldsymbol{\xi}' \boldsymbol{m}(\lambda_t; \boldsymbol{\theta}_0) \right\}^{-1/(1+\nu)} \right]^{-1}. \tag{2.20}$$

Put $Y_t = \boldsymbol{\xi}' \boldsymbol{m}(\lambda_t; \boldsymbol{\theta}_0)$ and write $\boldsymbol{\xi} = \|\boldsymbol{\xi}\| \boldsymbol{u}$ where $\boldsymbol{u} \in U$ is a unit vector. From the second condition of (2.18), we have

$$\frac{1}{n} \boldsymbol{u}' \sum_{t=1}^{n} p_t \boldsymbol{m}(\lambda_t; \boldsymbol{\theta}_0) = 0,$$

and by substituting (2.19) and (2.20) into this equation, we have

$$\begin{aligned} 0 &= \frac{1}{n} \boldsymbol{u}' \sum_{t=1}^{n} \boldsymbol{m}(\lambda_t; \boldsymbol{\theta}_0)(1 + Y_t)^{-1/(1+\nu)} \\ &= \frac{1}{n} \boldsymbol{u}' \sum_{t=1}^{n} \boldsymbol{m}(\lambda_t; \boldsymbol{\theta}_0) \left\{ 1 - \frac{(1 + \eta_t Y_t)^{-1/(1+\nu)-1}}{1+\nu} Y_t \right\} \quad (0 < \eta_t < 1). \end{aligned}$$

We rewrite this as

$$||\boldsymbol{\xi}|| = \left[\boldsymbol{u}'\left\{\frac{1}{n}\sum_{t=1}^{n}\{1+\eta_t||\boldsymbol{\xi}||\boldsymbol{u}'\boldsymbol{m}(\lambda_t;\boldsymbol{\theta}_0)\}^{-1/(1+\nu)-1}\boldsymbol{m}(\lambda_t;\boldsymbol{\theta}_0)\boldsymbol{m}(\lambda_t;\boldsymbol{\theta}_0)'\right\}\boldsymbol{u}\right]^{-1}$$
$$\times(1+\nu)\boldsymbol{u}'\left\{\frac{1}{n}\sum_{t=1}^{n}\boldsymbol{m}(\lambda_t;\boldsymbol{\theta}_0)\right\}. \tag{2.21}$$

From Lemmas 2.1–2.3, there exists a sufficiently large C and an integer n_0 such that if $n > n_0$ and $||\boldsymbol{\xi}|| < Cn^{-1/2}$, the right hand side of (2.21) is less than $Cn^{-1/2}$. Applying the Brouwer fixed point theorem to the right hand side of (2.21), we can see that a solution of Eq. (2.21) exists in the region $||\boldsymbol{\xi}|| < Cn^{-1/2}$. Hence it is shown that

$$||\boldsymbol{\xi}|| = O_p(n^{-1/2}). \tag{2.22}$$

Now, from the second condition of (2.18), we obtain

$$\begin{aligned}
\boldsymbol{0} &= \frac{1}{n}\sum_{t=1}^{n}\boldsymbol{m}(\lambda_t;\boldsymbol{\theta}_0)(1+Y_t)^{-1/(1+\nu)}\\
&= \frac{1}{n}\sum_{t=1}^{n}\boldsymbol{m}(\lambda_t;\boldsymbol{\theta}_0)\left\{1-(1+\nu)^{-1}Y_t+O_p(Y_t^2)\right\}\\
&= n^{-1/2}\boldsymbol{P}_n-(1+\nu)^{-1}\boldsymbol{S}_n\boldsymbol{\xi}+\boldsymbol{\beta}.
\end{aligned}$$

With some constant C, the norm of the last term $\boldsymbol{\beta}$ is evaluated as

$$||\boldsymbol{\beta}|| = \frac{C}{n}\sum_{t=1}^{n}||\boldsymbol{m}(\lambda_t;\boldsymbol{\theta}_0)||^3||\boldsymbol{\xi}||^2 \le CZ_n\boldsymbol{S}_n||\boldsymbol{\xi}||^2 = O(\log n)O_p(n^{-1})$$

and we have

$$\boldsymbol{\xi} = (1+\nu)n^{-1/2}\boldsymbol{S}_n^{-1}\boldsymbol{P}_n+\boldsymbol{\beta}.$$

Now it is shown that

$$\sum_{t=1}^{n}Y_t = \boldsymbol{\xi}'\sum_{t=1}^{n}\boldsymbol{m}(\lambda_t;\boldsymbol{\theta}_0) \;=\; (1+\nu)\boldsymbol{P}_n'\boldsymbol{S}_n^{-1}\boldsymbol{P}_n+O(\log n)O_p(n^{-1/2}), \tag{2.23}$$

$$\begin{aligned}
\sum_{t=1}^{n}Y_t^2 &= \boldsymbol{\xi}'\left(\sum_{t=1}^{n}\boldsymbol{m}(\lambda_t;\boldsymbol{\theta}_0)\boldsymbol{m}(\lambda_t;\boldsymbol{\theta}_0)'\right)\boldsymbol{\xi}\\
&= (1+\nu)^2\boldsymbol{P}_n'\boldsymbol{S}_n^{-1}\boldsymbol{P}_n+O(\log n)O_p(n^{-1/2})
\end{aligned} \tag{2.24}$$

$$\sum_{t=1}^{n} Y_t^3 \leq \max_{1\leq t\leq n} |Y_t| \sum_{t=1}^{n} Y_t^2 = ||\boldsymbol{\xi}|| \max_{1\leq t\leq n} ||\boldsymbol{m}(\lambda_t; \boldsymbol{\theta}_0)|| \sum_{t=1}^{n} Y_t^2$$
$$= O(\log n) O_p(n^{-1/2}). \tag{2.25}$$

From (2.19) and (2.20), we can write

$$\begin{aligned} CR_\nu(\boldsymbol{p}) &= \frac{2}{\nu(\nu+1)} \sum_{t=1}^{n} \left\{ (np_t)^{-\nu} - 1 \right\} \\ &= \frac{2}{\nu(\nu+1)} \sum_{t=1}^{n} \left[\left\{ \frac{n(1+Y_t)^{-1/(1+\nu)}}{\sum_{t=1}^{n}(1+Y_t)^{-1/(1+\nu)}} \right\}^{-\nu} - 1 \right] \\ &= \frac{2}{\nu(\nu+1)} \left[\left\{ \frac{1}{n} \sum_{t=1}^{n} (1+Y_t)^{-1/(1+\nu)} \right\}^{\nu} \left\{ \sum_{t=1}^{n} (1+Y_t)^{\nu/(1+\nu)} \right\} - n \right]. \end{aligned} \tag{2.26}$$

Taylor expansion leads to

$$\begin{aligned} &\left\{ \frac{1}{n} \sum_{t=1}^{n} (1+Y_t)^{-1/(1+\nu)} \right\}^{\nu} \\ &= \left\{ 1 - \frac{1}{n(1+\nu)} \sum_{t=1}^{n} Y_t + \frac{2+\nu}{2n(1+\nu)^2} \sum_{t=1}^{n} Y_t^2 + \frac{1}{n} \sum_{t=1}^{n} O_p(Y_t^3) \right\}^{\nu} \\ &= 1 - \frac{\nu}{n(1+\nu)} \sum_{t=1}^{n} Y_t + \frac{\nu(2+\nu)}{2n(1+\nu)^2} \sum_{t=1}^{n} Y_t^2 + \frac{1}{n} \sum_{t=1}^{n} O_p(Y_t^3) \; + \; O_p(n^{-2}) \end{aligned} \tag{2.27}$$

and

$$\sum_{t=1}^{n} (1+Y_t)^{\nu/(1+\nu)} = n + \frac{\nu}{1+\nu} \sum_{t=1}^{n} Y_t - \frac{\nu}{2(1+\nu)^2} \sum_{t=1}^{n} Y_t^2 + \sum_{t=1}^{n} O_p(Y_n^3). \tag{2.28}$$

Using (2.23)–(2.25), (2.27) and (2.28), we can show that

$$(\text{A.13}) = \frac{1}{(1+\nu)^2} \sum_{t=1}^{n} Y_t^2 + O(\log n) O_p(n^{-1/2}) = \boldsymbol{P}_n' \boldsymbol{S}_n^{-1} \boldsymbol{P}_n + O(\log n) O_p(n^{-1/2}).$$

Finally, from Lemmas 2.1 and 2.2, we can show that

$$\boldsymbol{P}_n' \boldsymbol{S}_n^{-1} \boldsymbol{P}_n \xrightarrow{d} \left(\boldsymbol{\Sigma}_2^{-1/2} \boldsymbol{\Sigma}_1^{1/2} \boldsymbol{\Sigma}_1^{-1/2} \boldsymbol{P}_n \right)' \left(\boldsymbol{\Sigma}_2^{-1/2} \boldsymbol{\Sigma}_1^{1/2} \boldsymbol{\Sigma}_1^{-1/2} \boldsymbol{P}_n \right) \stackrel{d}{=} (\boldsymbol{\Sigma} \boldsymbol{N})'(\boldsymbol{\Sigma} \boldsymbol{N})$$

This is the desired result.

When $\nu = 0$, Theorem 2.3 is reduced to Theorem 2.2. When $\nu = -1$, the proof is similar. □

References

Baggerly, K.A.: Empirical likelihood as a goodness-of-fit measure. Biometrika **85**, 535–547 (1998)

Belkacem, L., Véhel, J.L., Walter, C.: Capm, risk and portfolio selection in alpha-stable markets. Fractals **8**, 99–115 (2000)

Brillinger, D.R.: Time Series: Data Analysis and Theory, expanded edn. Holden-Day, San Francisco (2001)

Brockwell, P.J., Davis, R.A.: Time Series: Theory and Methods, 2nd edn. Springer, Berlin (1991)

Dominicy, Y., Veredas, D.: The method of simulated quantiles. J. Econometrics **172**, 235–247 (2013)

Embrechts, P., Klüppelberg, C., Mikosch, T.: Modelling Extreme Events for Insurance and Finance. Springer-Verlag, Berlin (1997)

Fama, E.F.: The behavior of stock-market prices. J. Bus. **38**, 34–105 (1965)

Hannan, E.J.: Multiple Time Series. John Wiley, New York (1970)

Hansen, L.P., Heaton, J., Yaron, A.: Finite-sample properties of some alternative GMM estimators. J. Bus. Econ. Statist. **14**, 262–280 (1996)

Kitamura, Y.: Empirical likelihood methods with weakly dependent processes. Ann. Statist. **25**, 2084–2102 (1997)

Kitamura, Y., Stutzer, M.: An information-theoretic alternative to generalized method of moments estimation. Econometrica **65**, 861–874 (1997)

Mandelbrot, B.: The variation of certain speculative prices. J. Bus. **36**, 394–419 (1963)

Monti, A.C.: Empirical likelihood confidence regions in time series models. Biometrika **80**, 329–338 (1997)

Nolan, J.P., Panorska, A.K., McCulloch, J.H.: Estimation of stable spectral measures. Math. Comput. Model. **34**, 1113–1122 (2001)

Nordman, D.J., Lahiri, S.N.: A frequency domain empirical likelihood for short- and long-range dependence. Ann. Statist. **34**, 3019–3050 (2006)

Newey, W.K., Smith, R.J.: Higher order properties of GMM and generalized empirical likelihood estimators. Econometrica **72**, 219–255 (2004)

Ogata, H.: Optimal portfolio estimation for dependent financial returns with generalized empirical likelihood. Adv. Decis. Sci. **2012**, 8pages, Article ID 973173 (2012).

Ogata, H.: Estimation for multivariate stable distributions with generalized empirical likelihood. J. Econometrics **172**, 248–254 (2013). doi:10.1016/j.jeconom.2012.08.017

Ogata, H., Taniguchi, M.: Cressie-Read power-divergence statistics for non Gaussian vector stationary processes. Scand. J. Stat. **36**, 141–156 (2009)

Ogata, H., Taniguchi, M.: An empirical likelihood approach for non-Gaussian vector stationary processes and its application to minimum contrast estimation. Aust. N. Z. J. Stat. **52**, 451–468 (2010)

Owen, A.B.: Empirical likelihood ratio confidence intervals for a single functional. Biometrika **75**, 237–249 (1988)

Owen, A.B.: Empirical likelihood confidence regions. Ann. Statist. **18**, 90–120 (1990)

Owen, A.B.: Empirical likelihood. Chapman & Hall/CRC, Boca Raton, London, New York, Washington, DC (2001)

Qin, J., Lawless, J.: Empirical likelihood and general estimating equations. Ann. Statist. **22**, 300–325 (1994)

Rachev, S.T., Han, S.: Portfolio management with stable distributions. Math. Method. Oper. Res. **51**, 341–352 (2000)

Read, T.R.C., Cressie, N.A.C.: Goodness-of-fit statisticsfor discrete multivariate data. Springer-Verlag, New York (1988)

Yu, J.: Empirical characteristic function estimation and its applications. Econometric Rev. **23**, 93–123 (2004)

Chapter 3
Various Methods for Financial Engineering

Abstract Various statistical methods have been introduced to many application fields. Such methods are often designed for standard settings, i.e., i.i.d. cases, regular model etc. However, financial data are usually dependent and have complicated features (see Chap. 1). In this chapter, we state various methods which are suitable for financial data. In Sect. 3.2, the control variate method is applied to time series models. Control variate method is the one to reduce the variance of estimators. However, this method has been developed mainly in i.i.d. cases. Because financial data are usually dependent, we extend this method to dependent case. In Sect. 3.3, we apply an instrumental variable method to a stochastic regression model. In stochastic regression models, a natural estimator for the regression coefficients is the ordinary least squares estimator (OLS). However, if the explanatory variable and the stochastic disturbance are correlated, this estimator is inconsistent. To overcome this difficulty, the instrumental variable method is used. In the CAPM model, it will be shown that the explanatory variable and the disturbance are fractionally cointegrated. Hence, we use the instrumental variable method to estimate the regression coefficients.

Keywords Control variate method · Instrumental variable method · Stochastic regression model · CAPM · Long memory process · Ordinary least squares estimator

3.1 Introduction

In this chapter, we state two important methods which are suitable for financial data.

In Sect. 3.2, the control variate method is applied to time series models. The sample mean is one of the most natural estimators of the population mean based on i.i.d. sample. However, if some control variables are available, it is known that the control variate method reduces the variance of the sample mean. This method has been discussed in the case when the sample and control variable are i.i.d.. Here we

M. Taniguchi et al., *Statistical Inference for Financial Engineering*, SpringerBriefs in Statistics, DOI: 10.1007/978-3-319-03497-3_3,

assume that these variables are stationary processes. Then we propose an estimator of the mean of the stationary process of interest by using control variate method. It is shown that this estimator improves the sample mean in the sense of mean square error. Also this analysis is extended to the case when the mean dynamics is of the form of regression.

In Sect. 3.3, we apply an instrumental variable method to the stochastic regression model, which includes CAPM model with time dimension. In the CAPM, empirical studies suggest that the response process and the explanatory process are short memory dependent and long memory dependent, respectively. From this point of view, we have to assume that the error process is also long memory dependent and is correlated with the explanatory process. For the stochastic regression model, the most fundamental estimator is the ordinary least squares estimator. However, the dependence of the error process with the explanatory process makes this estimator to be inconsistent. To overcome this difficulty instrumental variable method was proposed. In this section, by using instrumental variable method, we propose the two-stage least squares (2SLS) estimator for the stochastic regression model in which the explanatory process and error process are long memory dependent and mutually correlated each other. Then we prove its consistency and CLT under some conditions.

3.2 Control Variate Methods for Financial Data

When we deal with financial time series data, estimation of dynamics mean is most fundamental, and is useful for the prediction of financial time series and construction of portfolio on financial assets etc.

For this purpose, the sample mean is the most natural one. However, it is often observed that the sample mean is not so good. In such a situation, if we can use some control variate information, we can introduce more effective method to estimate the dynamics mean, which is called the control variate method.

When some control variable vector is available (a random vector which is possibly correlated with the variable of interest), using the information about the control variate vector, it is known that the control variate method reduces the variance of the sample mean. That is, if μ_Y is an unknown mean of i.i.d. data $\{Y(1), Y(2), \ldots, Y(n)\}$ and $\boldsymbol{X}$ is a control variable vector with known mean vector $\boldsymbol{\mu}_X$, then for any constant vector $\boldsymbol{b}$, the mean of the control variate estimator $\hat{\mu}_Y(\boldsymbol{b}) = \bar{Y}_n - \boldsymbol{b}'(\boldsymbol{X} - \boldsymbol{\mu}_X)$ is μ_Y ($\bar{Y}_n$ is the sample mean of $\{Y(1), Y(2), \ldots, Y(n)\}$) and its variance is $Var\{\bar{Y}_n\} - 2\boldsymbol{b}'Cov\{\bar{Y}_n, \boldsymbol{X}\} + \boldsymbol{b}'\sum_X \boldsymbol{b}$, where $\sum_X$ is the covariance matrix of $\boldsymbol{X}$ and $Cov\{\bar{Y}_n, \boldsymbol{X}\}$ is the covariance vector between $\bar{Y}_n$ and $\boldsymbol{X}$. Hence if $2\boldsymbol{b}'Cov\{\bar{Y}_n, \boldsymbol{X}\} > \boldsymbol{b}'\sum_X \boldsymbol{b}$, then the variance of the control variate estimator is smaller than that of the sample mean.

This method has been discussed in the case when the sample and control variable are i.i.d.. Lavenberg and Welch (1981) reviews analyses of the control variate developed up to the date. In the paper, the value $\boldsymbol{b}_*$ of vector $\boldsymbol{b}$ which minimizes the variance of the control variate estimator is derived and the confidence interval of

$\hat{\mu}_Y(\boldsymbol{b}_*)$ is constructed. However, in practice, since the correlation between $\bar{Y}_n$ and $\boldsymbol{X}$ is unknown, this $\boldsymbol{b}_*$ is not known and an estimator $\hat{\boldsymbol{b}}_*$ of $\boldsymbol{b}_*$ is proposed. In general, the control variate estimator involving the estimator $\hat{\boldsymbol{b}}_*$ is not unbaiased and the confidence interval cannot be constructed easily. They also discuss these problems. Rubinstein and Marcus (1985) extends the results to the case when the sample mean $\bar{\boldsymbol{Y}}_n$ is multidimensional vector and the multidimensional control variate estimator is represented as $\hat{\boldsymbol{\mu}}_Y(\boldsymbol{B}) = \bar{\boldsymbol{Y}}_n - \boldsymbol{B}(\boldsymbol{X} - \boldsymbol{\mu}_X)$, where $\boldsymbol{B}$ is an arbitrary matrix and $\boldsymbol{X}$ is a control variate vector with mean vector $\boldsymbol{\mu}_X$. Giving the matrix $\boldsymbol{B}_*$ which minimizes the determinant of $E\{\hat{\boldsymbol{\mu}}_Y(\boldsymbol{B})'\hat{\boldsymbol{\mu}}_Y(\boldsymbol{B})\}$, they introduce an estimator $\hat{\boldsymbol{\mu}}_Y(\boldsymbol{B}_*)$. They also introduce an estimator of $\hat{\boldsymbol{B}}_*$ of $\boldsymbol{B}_*$ and discuss the confidence ellipsoid. Nelson (1990) proves a central limit theorem of the control variate estimator. Since a lot of control variate theories have been discussed under a specific probability structure (usually normal distribution) for the sample and control variates, a number of authors introduced remedies for violations of these assumptions. Nelson (1990) gives a systematic analytical evaluation of them. In recent years, this method has been applied to financial engineering (e.g., Glasserman (2004), Chan and Wong (2006)).

As we saw in Chap. 1, empirical studies show that many financial data are dependent. However, the control variate theory is usually discussed under the assumption that the sample and control variates are i.i.d. Hence in this section, when the sample is generated from a stationary process and some control variable process is available, we propose an estimator $\hat{\theta}_C$ of the mean of the concerned process by using control variate method. Then it is shown that this estimator improves the sample mean in the sense of mean square error (MSE). The estimator $\hat{\theta}_C$ is expressed in terms of nonparametric estimators for spectra of the concerned process and the control variate process. We also apply this analysis to the case when the mean dynamics is of the form of regression. A control variate estimator for the regression coefficients is proposed and is shown to improve the LSE in the sense of MSE. Numerical studies show how our estimators behave. Our results have applications to various fields, including finance in particular.

Suppose that $\{Y(t); t \in \mathbb{Z}\}$ is a scalar-valued process with mean $E\{Y(t)\} = \theta$ and $\{\boldsymbol{X}(t); t \in \mathbb{Z}\}$ is an another m-dimensional process with the mean vector $E\{\boldsymbol{X}(t)\} = \boldsymbol{0}$, which is possibly correlated with $\{Y(t)\}$. We are now interested in estimation of θ. Let $\boldsymbol{Z}(t) \equiv (Y(t), \boldsymbol{X}'(t))'$ and we assume $\{\boldsymbol{Z}(t); t \in \mathbb{Z}\}$ is generated by the following linear process.

$$\boldsymbol{Z}(t) = \sum_{j=0}^{\infty} \boldsymbol{B}(j)\boldsymbol{\epsilon}(t-j) + \boldsymbol{\theta} \tag{3.1}$$

where $\boldsymbol{\theta} = (\theta, 0, 0, \ldots, 0)'$ is $m+1$-dimensional vector and $\boldsymbol{B}(j)'s$ are $(m+1) \times (m+1)$ matrices, and $\{\boldsymbol{\epsilon}(t)\}$ is a sequence of i.i.d. $(m+1)$-dimensional random vectors with mean vector $\boldsymbol{0}$ and covariance matrix $\boldsymbol{K}$.

Henceforth $|\boldsymbol{U}|$ denotes the sum of all the absolute values of elements of matrix $\boldsymbol{U}$.

Assumption 3.1 (i) $det\{\sum_{u=0}^{\infty} \boldsymbol{B}(u)z^u\} = 0$ has no roots in the unit disc $\{z \in \mathbb{C}; |z| \leq 1\}$.
(ii) The coefficient matrices $\boldsymbol{B}(u)$ satisfy

$$\sum_{u=0}^{\infty} |u|^4 |\boldsymbol{B}(u)| < \infty. \tag{3.2}$$

Using the same k-th order cumulant notation as in Sect. 1.1, we set the following.

Assumption 3.2 For $k = 3, 4, \ldots,$

$$C_k^{\epsilon} \equiv \sup_{a_1, \ldots, a_k} \left| c_{a_1, \ldots, a_k}^{\epsilon}(0, \ldots, 0) \right| < \infty \tag{3.3}$$

and

$$\sum_{L=1}^{\infty} \left(\sum_{\nu} C_{n_1}^{\epsilon} \cdots C_{n_P}^{\epsilon} \right) z^L / L! < \infty, \tag{3.4}$$

for z in a neighborhood of 0, where the inner summation is over all indecomposable partitions (see p. 20 of Brillinger (2001)) $\nu = (\nu_1, \ldots, \nu_P)$ of the table

$$\begin{array}{cc} 1 & 2 \\ 3 & 4 \\ \vdots & \vdots \\ 2L-1 & 2L \end{array} \tag{3.5}$$

with ν_p having $n_p > 1$ elements, $p = 1, \ldots, P$.

Write

$$C_k \equiv \sup_{a_1, \ldots, a_k} \sum_{t_1, \ldots, t_{k-1} = -\infty}^{\infty} \left| c_{a_1, \ldots, a_k}^{z}(t_1, \ldots, t_{k-1}) \right|. \tag{3.6}$$

Then Assumptions 3.1 and 3.2 imply Assumption (**B**) in Chap. 1 and

$$\sum_{L=1}^{\infty} \left(\sum_{\nu} C_{n_1} \cdots C_{n_P} \right) z^L / L! < \infty, \tag{3.7}$$

in a neighborhood of 0, where the summation $\sum_{\nu}$ is defined as in (3.4) (see p. 48 of Brillinger (2001)). From Assumption 3.1, it is seen that the process $\{\boldsymbol{Z}(t)\}$ becomes a stationary process with nonsingular spectral density matrix (e.g., Brillinger (2001)). We write the spectral density matrix by

$$f(\lambda) = \begin{bmatrix} f_{YY}(\lambda) & \boldsymbol{f}_{YX}(\lambda) \\ \boldsymbol{f}_{XY}(\lambda) & \boldsymbol{f}_{XX}(\lambda) \end{bmatrix}. \tag{3.8}$$

From Assumption 3.1, it follows that $\boldsymbol{R}(s) = \{c^z_{i,j}(s);\, i,j = 1, \ldots, m+1\}$ satisfies

$$\sum_{s=-\infty}^{\infty} |s|^4 |\boldsymbol{R}(s)| < \infty, \tag{3.9}$$

(e.g., p. 46 of Brillinger (2001)). Suppose that partial observations $\{Y(0), Y(1), \ldots Y(n-1)\}$ and $\{\boldsymbol{X}(-M_n), \boldsymbol{X}(-M_n+1), \ldots, \boldsymbol{X}(0), \ldots, \boldsymbol{X}(n-1)\}$ are available, where $M_n = O(n^{\beta})$ $(\frac{1}{4} \le \beta < \frac{1}{3})$.

Now we are interested in the estimation of θ. Based on the observations, we introduce the following estimator $\hat{\theta}_C$ of θ

$$\hat{\theta}_C \equiv \frac{1}{n} \sum_{t=0}^{n-1} \left\{ Y(t) - \sum_{u=0}^{M_n} \hat{\boldsymbol{a}}'_n(u) \boldsymbol{X}(t-u) \right\}, \tag{3.10}$$

where $\hat{\boldsymbol{a}}_n(u) = \frac{1}{2\pi} \int_{-\pi}^{\pi} \hat{\boldsymbol{A}}_n(\lambda) \exp(iu\lambda) d\lambda$, $\hat{\boldsymbol{A}}_n(\lambda) = \hat{\boldsymbol{f}}_{XX}(\lambda)^{-1} \hat{\boldsymbol{f}}_{XY}(\lambda)$. Here $\hat{\boldsymbol{f}}_{XX}(\lambda)$ and $\hat{\boldsymbol{f}}_{XY}(\lambda)$ are, respectively, nonparametric estimators of $\boldsymbol{f}_{XX}(\lambda)$ and $\boldsymbol{f}_{XY}(\lambda)$ which are defined as,

$$\hat{\boldsymbol{f}}_{XY}(\lambda) \equiv \frac{2\pi}{n} \sum_{s=1}^{n-1} W_n\left(\lambda - \frac{2\pi s}{n}\right) \boldsymbol{I}_{XY}\left(\frac{2\pi s}{n}\right) \tag{3.11}$$

$$\hat{\boldsymbol{f}}_{XX}(\lambda) \equiv \frac{2\pi}{n} \sum_{s=1}^{n-1} W_n\left(\lambda - \frac{2\pi s}{n}\right) \boldsymbol{I}_{XX}\left(\frac{2\pi s}{n}\right) \tag{3.12}$$

where $\boldsymbol{I}_{XY}(\mu)$ and $\boldsymbol{I}_{XX}(\mu)$ are submatrices of the periodogram of $\boldsymbol{Z}(t)$

$$\boldsymbol{I}_n(\mu) = \begin{bmatrix} I_{YY}(\mu) & \boldsymbol{I}_{YX}(\mu) \\ \boldsymbol{I}_{XY}(\mu) & \boldsymbol{I}_{XX}(\mu) \end{bmatrix}, \tag{3.13}$$

and $\{W_n(\lambda)\}$ are weight functions which are described below. The $\hat{\boldsymbol{A}}_n(\lambda)$ and $\hat{\boldsymbol{a}}_n(u)$ are shown to be consistent estimators of $\boldsymbol{A}(\lambda) = \boldsymbol{f}_{XX}(\lambda)^{-1} \boldsymbol{f}_{XY}(\lambda)$, $\boldsymbol{a}(u) = \frac{1}{2\pi} \int_{-\pi}^{\pi} \boldsymbol{A}(\lambda) \exp(iu\lambda) d\lambda$, respectively.

Next, we will show that the proposed estimator $\hat{\theta}_C$ improves the sample mean in the sense of MSE.

Then we get the following theorem. For the proof, see Amano and Taniguchi (2011).

Theorem 3.1 *Suppose Assumptions* 3.1 *and* 3.2*, and that the spectral window* $W_n(\lambda)$ *satisfies Assumption* 1.2*. Then it holds that*

$$\lim_{n\to\infty} nE|\hat{\theta}_C - \theta|^2 = 2\pi \left(f_{YY}(0) - \boldsymbol{f}_{YX}(0)\boldsymbol{f}_{XX}(0)^{-1}\boldsymbol{f}_{XY}(0)\right). \tag{3.14}$$

It is known that the asymptotic variance of the sample mean $\bar{Y}_n \equiv \frac{1}{n}\sum_{t=0}^{n-1} Y(t)$ is $2\pi f_{YY}(0)$ (e.g., Theorem 5.2.1 of Brillinger (2001)). Since

$$2\pi \left(f_{YY}(0) - \boldsymbol{f}_{YX}(0)\boldsymbol{f}_{XX}(0)^{-1}\boldsymbol{f}_{XY}(0)\right) \le 2\pi f_{YY}(0), \tag{3.15}$$

we observe that $\hat{\theta}_C$ improves $\bar{Y}_n$ in the sense of MSE.

Next, we assume $\{Y(t);\, t \in \mathbb{Z}\}$ is a trend model whose mean $E[Y(t)] = \mu(t) = \boldsymbol{\phi}'(t)\boldsymbol{\theta}$ is a time-dependent function. Here $\boldsymbol{\phi}(t) = (\phi_1(t), \ldots, \phi_J(t))'$ and $\boldsymbol{\theta} = (\theta_1, \ldots, \theta_J)'$. Let $\{\boldsymbol{X}(t);\, t \in \mathbb{Z}\}$ be another m-dimensional process with mean vector $E\{\boldsymbol{X}(t)\} = \boldsymbol{0}$, which is possibly correlated with $\{Y(t)\}$. Now we apply the control variate method to estimate the parameter $\boldsymbol{\theta}$. Let $\boldsymbol{Z}(t) \equiv (Y(t), \boldsymbol{X}'(t))'$, $t \in \mathbb{Z}$. We impose the following assumption.

Assumption 3.3 $\{\boldsymbol{Z}(t);\, t \in \mathbb{Z}\}$ is generated by the following linear process.

$$\boldsymbol{Z}(t) = \sum_{j=0}^{\infty} \boldsymbol{B}(j)\boldsymbol{\epsilon}(t-j) + \begin{pmatrix} \mu(t) \\ 0 \\ \vdots \\ 0 \end{pmatrix} \tag{3.16}$$

where $\boldsymbol{B}(j)'s$ are $(m+1) \times (m+1)$ matrices satisfying Assumption 3.1 and $\boldsymbol{\epsilon}(t)'s$ are i.i.d. random vectors with mean vector $\boldsymbol{0}$ and covariance matrix $\boldsymbol{K}$.

For convenience, we define $\eta(t)$ by $\sum_{j=0}^{\infty} \boldsymbol{B}(j)\boldsymbol{\epsilon}(t-j) = (\eta(t), \boldsymbol{X}'(t))'$, then as discussed before, $(\eta(t), \boldsymbol{X}'(t))'$ has the spectral density matrix,

$$\boldsymbol{f}(\lambda) = \begin{bmatrix} f_{\eta\eta}(\lambda) & \boldsymbol{f}_{\eta X}(\lambda) \\ \boldsymbol{f}_{X\eta}(\lambda) & \boldsymbol{f}_{XX}(\lambda) \end{bmatrix}. \tag{3.17}$$

Suppose that partial observations $\{Y(0), Y(1), \ldots Y(n-1)\}$ and $\{\boldsymbol{X}(-M_n), \boldsymbol{X}(-M_n+1), \ldots, \boldsymbol{X}(0), \ldots, \boldsymbol{X}(n-1)\}$ are available.

We define nonparametric estimators $\hat{\boldsymbol{f}}_{XX}(\lambda)$ and $\hat{\boldsymbol{f}}_{X\hat{\eta}}(\lambda)$ for the spectral densities $\boldsymbol{f}_{XX}(\lambda)$ and $\boldsymbol{f}_{X\eta}(\lambda)$, respectively, as

$$\hat{\boldsymbol{f}}_{XX}(\lambda) \equiv \frac{2\pi}{n}\sum_{s=1}^{n-1} W_n\left(\lambda - \frac{2\pi s}{n}\right)\boldsymbol{I}_{XX}\left(\frac{2\pi s}{n}\right) \tag{3.18}$$

$$\hat{\boldsymbol{f}}_{X\hat{\eta}}(\lambda) \equiv \frac{2\pi}{n}\sum_{s=1}^{n-1} W_n\left(\lambda - \frac{2\pi s}{n}\right)\boldsymbol{I}_{X\hat{\eta}}\left(\frac{2\pi s}{n}\right) \tag{3.19}$$

where

$$\boldsymbol{I}_{\boldsymbol{XX}}(\mu) \equiv \frac{1}{2\pi n}\left\{\sum_{t=0}^{n-1}\boldsymbol{X}(t)e^{it\mu}\right\}\left\{\sum_{t=0}^{n-1}\boldsymbol{X}(t)e^{it\mu}\right\}^* \tag{3.20}$$

$$\boldsymbol{I}_{\boldsymbol{X}\hat{\eta}}(\mu) \equiv \frac{1}{2\pi n}\left\{\sum_{t=0}^{n-1}\boldsymbol{X}(t)e^{it\mu}\right\}\left\{\sum_{t=0}^{n-1}\hat{\eta}(t)e^{it\mu}\right\}^* \tag{3.21}$$

where $\hat{\eta}(t) = Y(t) - \boldsymbol{\phi}'(t)\hat{\boldsymbol{\theta}}_{LSE}$, $\hat{\boldsymbol{\theta}}_{LSE} = (\boldsymbol{\phi}'\boldsymbol{\phi})^{-1}\boldsymbol{\phi}'\boldsymbol{Y}$ (the least squares estimator of $\boldsymbol{\theta}$), $\boldsymbol{Y} = (Y(1), \ldots, Y(n))'$ and $\boldsymbol{\phi} = (\boldsymbol{\phi}(1), \ldots, \boldsymbol{\phi}(n))'$. Let $\hat{\boldsymbol{A}}(\lambda) = \hat{\boldsymbol{f}}_{\boldsymbol{XX}}(\lambda)^{-1}\hat{\boldsymbol{f}}_{\boldsymbol{X}\hat{\eta}}(\lambda)$ and $\hat{\boldsymbol{a}}(u) = \frac{1}{2\pi}\int_{-\pi}^{\pi}\hat{\boldsymbol{A}}(\lambda)\exp(iu\lambda)d\lambda$.

Now we propose an estimator $\hat{\boldsymbol{\theta}}_{LSE}^{C}$ of $\boldsymbol{\theta}$:

$$\hat{\boldsymbol{\theta}}_{LSE}^{C} = (\boldsymbol{\phi}'\boldsymbol{\phi})^{-1}\boldsymbol{\phi}'(\boldsymbol{Y} - \hat{\boldsymbol{W}}_M) \tag{3.22}$$

where $\hat{\boldsymbol{W}}_M = (\hat{W}_M(1), \ldots, \hat{W}_M(n))'$ with

$$\hat{W}_M(t) = \sum_{u=0}^{M_n}\hat{\boldsymbol{a}}'(u)\boldsymbol{X}(t-u). \tag{3.23}$$

To describe the asymptotics of $\hat{\boldsymbol{\theta}}_{LSE}^{C}$, we impose the following Grenander's conditions.

Assumption 3.4 Let $c_{j,k}^n(h) = \sum_{t=1}^{n-h}\phi_j(t+h)\phi_k(t) = \sum_{t=1-h}^{n}\phi_j(t+h)\phi_k(t)$. $c_{j,k}^n(h)$'s satisfy

(i) $c_{j,j}^n(0) = O(n^\gamma), j = 1, \ldots, J$ for some $\gamma > 0$.

(ii) $\lim_{n\to\infty}\frac{\phi_j^2(n+1)}{c_{j,j}^n(0)} = 0, j = 1, \ldots, J$.

(iii)

$$\lim_{n\to\infty}\frac{c_{j,k}^n(h)}{\left\{c_{j,j}^n(0)c_{k,k}^n(0)\right\}^{\frac{1}{2}}} = m_{jk}(h) \tag{3.24}$$

We may take $\phi_1(t) = 1$ (constant), which evidently satisfies Assumption 3.4, hence, the regression part $\boldsymbol{\phi}(t)$ of $\{Y(t)\}$ may include a constant.

We define the $J \times J$ matrix $\boldsymbol{m}_{\boldsymbol{\phi\phi}}(u)$ by

$$\boldsymbol{m}_{\boldsymbol{\phi\phi}}(u) = \{m_{jk}(u); j, k = 1, \ldots, J\}. \tag{3.25}$$

From p. 175 of Brillinger (2001), there exists an $r \times r$ matrix-valued function $\boldsymbol{G}_{\boldsymbol{\phi\phi}}(\lambda)$, $-\pi < \lambda \leq \pi$, whose entries are of bounded variation, such that

$$\boldsymbol{m}_{\boldsymbol{\phi\phi}}(u) = \int_{-\pi}^{\pi} \exp(iu\lambda) d\boldsymbol{G}_{\boldsymbol{\phi\phi}}(\lambda) \tag{3.26}$$

for $u = 0, \pm 1, \cdots$ Under these assumptions, we obtain the following theorem. For the proof, see Amano and Taniguchi (2011).

Theorem 3.2 *Suppose Assumptions* 3.2–3.4, *then*

$$\lim_{n\to\infty} n^{\gamma} E\{(\hat{\boldsymbol{\theta}}_{LSE}^{C} - \boldsymbol{\theta})(\hat{\boldsymbol{\theta}}_{LSE}^{C} - \boldsymbol{\theta})'\}$$
$$= 2\pi \boldsymbol{m}_{\boldsymbol{\phi\phi}}(0)^{-1} \int_{-\pi}^{\pi} f_{\eta-V,\eta-V}(\lambda) d\boldsymbol{G}_{\boldsymbol{\phi\phi}}(\lambda) \boldsymbol{m}_{\boldsymbol{\phi\phi}}(0)^{-1}, \tag{3.27}$$

where $f_{\eta-V,\eta-V}(\lambda) = f_{\eta,\eta}(\lambda) - \boldsymbol{f}_{\eta \boldsymbol{X}}(\lambda) \boldsymbol{f}_{\boldsymbol{XX}}(\lambda)^{-1} \boldsymbol{f}_{\boldsymbol{X}\eta}(\lambda)$ *is the spectral density of* $\eta(t) - V(t)$. *Here* $V(t) = \sum_{u=0}^{\infty} \boldsymbol{a}'(u) \boldsymbol{X}(t-u)$, $\boldsymbol{a}(u) = \frac{1}{2\pi} \int_{-\pi}^{\pi} \boldsymbol{A}(\lambda) \exp(iu\lambda) d\lambda$, $\boldsymbol{A}(\lambda) = \boldsymbol{f}_{\boldsymbol{XX}}(\lambda)^{-1} \boldsymbol{f}_{\boldsymbol{X}\eta}(\lambda)$.

Note that the least squares estimator $\hat{\boldsymbol{\theta}}_{LSE}$ of $\boldsymbol{\theta}$ has the following asymptotic variance

$$\lim_{n\to\infty} n^{\gamma} E\{(\hat{\boldsymbol{\theta}}_{LSE} - \boldsymbol{\theta})(\hat{\boldsymbol{\theta}}_{LSE} - \boldsymbol{\theta})'\}$$
$$= 2\pi \boldsymbol{m}_{\boldsymbol{\phi\phi}}(0)^{-1} \int_{-\pi}^{\pi} f_{\eta,\eta}(\lambda) d\boldsymbol{G}_{\boldsymbol{\phi\phi}}(\lambda) \boldsymbol{m}_{\boldsymbol{\phi\phi}}(0)^{-1}, \tag{3.28}$$

where $f_{\eta,\eta}(\lambda)$ is the spectral density of $\eta(t)$. It is seen that

$$f_{\eta-v,\eta-v}(\lambda) \equiv f_{\eta,\eta}(\lambda) - \boldsymbol{f}_{\eta \boldsymbol{X}}(\lambda) \boldsymbol{f}_{\boldsymbol{XX}}(\lambda)^{-1} \boldsymbol{f}_{\boldsymbol{X}\eta}(\lambda) \leq f_{\eta,\eta}(\lambda), \tag{3.29}$$

which implies that the asymptotic covariance matrix of $\hat{\boldsymbol{\theta}}_{LSE}^{C}$ is smaller than that of $\hat{\boldsymbol{\theta}}_{LSE}$.

Next we examine the control variate estimators numerically. In Example 3.1, we compare the control variate estimators with sample means. Next the control variate estimators are compared with the least squares estimators in Example 3.2. In Example 3.3, we investigate its usefulness in real financial data.

Example 3.1 We consider the following model,

$$Y(t) = u(t) + v(t) \tag{3.30}$$
$$X(t) = a_1 u(t) + 0.4u(t-1) + a_2 v(t), \tag{3.31}$$

where $\{u(t)\}$, $\{v(t)\} \sim$ i.i.d. $N(0,1)$ and they are mutually independent. The length of $Y(t)$ and $X(t)$ are set by 1,000 and based on 5,000 times simulation we report MSE of $\hat{\theta}_C$ with $M_n = 20$ and $\bar{Y}_n$. We set $a_1 = 0.1, 0.2, 0.3$ and $a_2 = 0.1, 0.2, 0.3, 4.0$ in Table 3.1.

Table 3.1 MSE of $\hat{\theta}_C$ and $\bar{Y}_n$

a_1	a_2	MSE $\hat{\theta}_C$	MSE $\bar{Y}_n$	MSE $\bar{Y}_n$ − MSE $\hat{\theta}_C$
0.1	0.1	0.00062385	0.00205319	0.00142934
0.1	0.2	0.00030573	0.00196825	0.00166252
0.1	0.3	0.00012184	0.00201872	0.00189688
0.2	0.1	0.00068064	0.00193495	0.00125431
0.2	0.2	0.00042055	0.00206655	0.001646
0.2	0.3	0.0002056	0.00192683	0.00172123
0.3	0.1	0.00071185	0.0020196	0.00130775
0.3	0.2	0.00047111	0.00195619	0.00148508
0.3	0.3	0.00027794	0.00198016	0.00170222
0.3	4.0	0.00068665	0.00199143	0.00130478

Table 3.2 MSE of $\hat{\boldsymbol{\theta}}^C_{LSE}$ and $\hat{\boldsymbol{\theta}}_{LSE}$ for $\boldsymbol{\phi}(t) = (1, t)'$

a_2	MSE $\hat{\boldsymbol{\theta}}^C_{LSE}$	MSE $\hat{\boldsymbol{\theta}}_{LSE}$	MSE $\hat{\boldsymbol{\theta}}_{LSE}$ − MSE $\hat{\boldsymbol{\theta}}^C_{LSE}$
0.1	0.00786141	0.01047323	0.00261182
0.2	0.0041377	0.00543288	0.00129518
0.3	0.00493462	0.00533173	0.00039711

From Table 3.1, we can see MSE$\bar{Y}_n$ − MSE$\hat{\theta}_C$ becomes larger as a_2 becomes large, which implies, if control variates are highly correlated with the disturbance, then $\hat{\theta}_C$ is better than $\bar{Y}_n$. However, excessive influence of the disturbance makes the performance of $\hat{\theta}_C$ worse.

Example 3.2 Consider the following model,

$$Y(t) = \mu(t) + u(t) + v(t) \tag{3.32}$$

$$X(t) = 0.3u(t) + 0.4u(t-1) + a_2 v(t), \tag{3.33}$$

where $\{u(t)\}, \{v(t)\} \sim$ i.i.d. $N(0, 1)$ and they are mutually independent. Here $\mu(t) = \boldsymbol{\theta}'\boldsymbol{\phi}(t)$, $\boldsymbol{\phi}(t)$ is a regression function and $\boldsymbol{\theta}$ is a vector-valued parameter. The length of $Y(t)$ and $X(t)$ are set by 1,000 and based on 5,000 replications we report MSE of $\hat{\boldsymbol{\theta}}^C_{LSE}$ with $M_n = 20$ and $\hat{\boldsymbol{\theta}}_{LSE}$, that is MSE of $\hat{\boldsymbol{\theta}}^C_{LSE}$ and $\hat{\boldsymbol{\theta}}_{LSE}$ are $\frac{1}{5{,}000}\sum_{i=1}^{5{,}000} ||\hat{\boldsymbol{\theta}}^C_{LSE}(i) - \boldsymbol{\theta}||^2$ and $\frac{1}{5{,}000}\sum_{i=1}^{5{,}000} ||\hat{\boldsymbol{\theta}}_{LSE}(i) - \boldsymbol{\theta}||^2$ ($\hat{\boldsymbol{\theta}}^C_{LSE}(i)$ is ith control variate estimator and $\hat{\boldsymbol{\theta}}_{LSE}(i)$ is ith least squares estimator). We set $\boldsymbol{\theta} = (1, 1)'$ and $a_2 = 0.1, 0.2, 0.3$. Table 3.2 shows MSE of $\hat{\boldsymbol{\theta}}^C_{LSE}$ and $\hat{\boldsymbol{\theta}}_{LSE}$ for $\boldsymbol{\phi}(t) = (1, t)'$, and Table 3.3 shows those for $\boldsymbol{\phi}(t) = (1, cos(\frac{\pi}{4}t))'$.

From Tables 3.2 and 3.3, we observe that MSE $\hat{\boldsymbol{\theta}}^C_{LSE}$ are smaller than MSE $\hat{\boldsymbol{\theta}}_{LSE}$. That is, control variate estimator also improves the least squares estimator.

Table 3.3 MSE of $\hat{\boldsymbol{\theta}}^C_{LSE}$ and $\hat{\boldsymbol{\theta}}_{LSE}$ for $\boldsymbol{\phi}(t) = (1, cos(\frac{\pi}{4}t))'$

a_2	MSE $\hat{\boldsymbol{\theta}}^C_{LSE}$	MSE $\hat{\boldsymbol{\theta}}_{LSE}$	MSE $\hat{\boldsymbol{\theta}}_{LSE}$ − MSE $\hat{\boldsymbol{\theta}}^C_{LSE}$
0.1	0.00594558	0.00837971	0.00243413
0.2	0.00747352	0.00749953	0.00002601
0.3	0.00913403	0.00940378	0.00026975

Taken from Amano and Taniguchi (2011). Published with the kind permission of © Elsevier B.V. 2011. All Rights Reserved

Table 3.4 Prediction of the stock price $S(N+H)$ by $\hat{\theta}_{H,C}$ and $\bar{Y}_{H,n}$

H	$\hat{S}^C(N+H)$	$\hat{S}(N+H)$	$S(N+H)$
1	891.0968	891.4768	891
2	888.3209	888.7471	876

Modified from Amano and Taniguchi (2011). Published with the kind permission of © Elsevier B.V. 2011. All Rights Reserved

Example 3.3 We calculate the control variate estimator $\hat{\theta}_{H,C}$ and the sample mean $\bar{Y}_{H,n}$ of $\{Y_H(t)\}$ where $Y_H(t) = \log S(t) - \log S(t-H)$ and $\{S(t)\}$ are NIPPON OIL CORPORATION's stock prices (7/20/2007 ∼ 12/12/2007). We set the difference between Yen–Euro's exchange rate and its sample mean (7/4/2007 ∼ 12/11/2007) as the control variate process. Then by use of $\hat{\theta}_{H,C}$ and $\bar{Y}_{H,n}$, we forecast NIPPON OIL CORPORATION's stock $S(N+H)$ at N = 12/12/2007 by $\hat{S}^C(N+H) \equiv e^{\hat{\theta}_{H,C}+\log S(N)}$ and $\hat{S}(N+H) \equiv e^{\bar{Y}_{H,n}+\log S(N)}$ in Table 3.4.

From Table 3.4, the prediction values $\hat{S}^C(N+H)$ are nearer to the true values $S(N+H)$ than $\hat{S}(N+H)$, which implies the prediction by the control variate estimator is better than that by the sample mean.

There are many cases in finance where we should estimate the statistical models for data of interest under the circumstance that we can use some related variables. In such situations, the estimators $\hat{\boldsymbol{\theta}}_C$ and $\hat{\boldsymbol{\theta}}^C_{LSE}$ can be used, and are more efficient than the usual estimators.

3.3 Statistical Estimation for Stochastic Regression Models with Long Memory Dependence

The CAPM is one of the typical models of risk asset's price on equilibrium market and has been used for pricing individual stocks and portfolios. At first, Markowitz (1991) worked the groundwork of this model. In his research, he cast the investor's portfolio selection problem in terms of expected return and variance. Sharpe (1964) and Lintner (1965) developed Markowitz's idea for economical implication. Black (1972) derived a more general version of the CAPM. In their version, the CAPM is constructed based on the excess of the return of the asset over zero-beta return $E\{R_i\} = E\{R_{0m}\} + \beta_{im}(E\{R_m\} - E\{R_{0m}\})$, where R_i and R_m are the return of the ith asset and the market portfolio, and R_{0m} is the return of zero-beta portfolio of the market portfolio. Campbell et al. (1997) discussed the estimation of CAPM, but in their work they did not discuss the time dimension. However, in the econometric analysis, CAPM with the time dimension should be investigated, that is, the model should be represented as $R_i(t) = \alpha_{im} + \beta_{im} R_m(t) + \epsilon_i(t)$. These models are included in the stochastic regression model, $R_i(t) = \alpha_i + \boldsymbol{\beta}_i' \boldsymbol{Z}(t) + \epsilon_i(t)$, where $\boldsymbol{Z}(t)$ is the risk factors. In the CAPM with the time dimension, the risk is only the return of the market portfolio. However, there are some evidences of other common risk factors besides the return of the market portfolio and stochastic regression models allow more factors than the return of the maket portfolio as the risk. Hence here, we investigate the stochastic regression models. Recently from the empirical analysis, it is known that the return of asset follows a short memory process and if we assume $\boldsymbol{Z}(t)$ follows short memory process, such a stochastic regression model has been investigated enough. Besides, if we consider the volatility of the return of the market portfolio as the risk factor, $Z(t)$ follows long memory dependent. From this point of view, we observe that $Z(t)$ and the error process $\boldsymbol{\epsilon}(t)$ are long memory dependent and correlated with each other.

For the stochastic regression model, the most fundamental estimator is the ordinary least squares estimator. However, the dependence of the error process with the explanatory process makes this estimator to be inconsistent. To overcome this difficulty, the instrumental variable method is proposed by use of the instrumental variables which are uncorrelated with the error process and correlated with the explanatory process. This method was first used by Wright (1928) and many researchers developed this method (see, Reiersöl (1945) and Geary (1949) etc.). Comprehensive reviews are seen in White (2001). However, the instrumental variable method has been discussed in the case where the error process does not follow the long memory process and this makes the estimation difficult.

For the analysis of long memory process, Robinson and Hidalgo (1997) considered a stochastic regression model defined by $y(t) = \alpha + \boldsymbol{\beta}' \boldsymbol{x}(t) + u(t)$, where α, $\boldsymbol{\beta} = (\beta_1, \ldots, \beta_K)'$ are unknown parameters and the K-vector processes $\{\boldsymbol{x}(t)\}$ and $\{u(t)\}$ are long memory dependent with $E\{\boldsymbol{x}(t)\} = \boldsymbol{0}$, $E\{u(t)\} = 0$. Furthermore, in Choy and Taniguchi (2001), they consider the stochastic regression model $y(t) = \beta x(t) + u(t)$, where $\{x(t)\}$ and $\{u(t)\}$ are stationary process with $E\{x(t)\} = \mu \neq 0$, and Choy and Taniguchi (2001) introduced a ratio estimator, the least squares estimator, and the best linear unbiased estimator for β. However, Robinson and Hidalgo (1997) and Choy and Taniguchi (2001) assume that the explanatory process $\{\boldsymbol{x}(t)\}$ and the error process $\{u(t)\}$ are independent.

In this section, by using instrumental variable method we propose the two-stage least squares (2SLS) estimator for the stochastic regression models in which stochastic explanatory process and error process are long memory dependent and mutually correlated each other. Then we prove its consistency and CLT under some conditions. Also, some numerical studies are provided.

For Sharpe and Lintner version of CAPM (see Sharpe (1964) and Lintner (1965)), the expected return of asset i is given by

$$E\{R_i\} = R_f + \beta_{im} \left(E\{R_m - R_f\} \right), \tag{3.34}$$

where

$$\beta_{im} = \frac{Cov\{R_i, R_m\}}{Var\{R_m\}},$$

R_m is the return of the market portfolio and R_f is the return of the risk-free asset. Another Sharpe–Lintner CAPM (see Sharpe (1964) and Lintner (1965)) is defined for $Z_i \equiv R_i - R_f$,

$$E\{Z_i\} = \beta_{im} E\{Z_m\}$$

where

$$\beta_{im} = \frac{Cov\{Z_i, Z_m\}}{Var\{Z_m\}}$$

and $Z_m = R_m - R_f$.

Black (1972) derived a more general version of CAPM, which is written as

$$E\{R_i\} = \alpha_{im} + \beta_{im} E\{R_m\}, \tag{3.35}$$

where $\alpha_{im} = E\{R_{0m}\}(1 - \beta_{im})$ and R_{0m} is the return on the *zero-beta portfolio*.

Since CAPM is single-period model, (3.34) and (3.35) do not have a time dimension. However for econometric analysis of the model, it is necessary to add assumptions concerning the time dimension. Hence it is natural to consider the model with the time dimension

$$R_i(t) = \alpha_{im} + \beta_{im} R_m(t) + \epsilon_i(t)$$

where i denotes the asset, t denotes the period, and $R_i(t)$ and $R_m(t)$, $i = 1, \ldots, q$; $t = 1, \ldots, n$ are, respectively, considered as the returns of the asset i and the market portfolio at t.

Indeed, there are some evidences of other common risk factors besides the return of market portfolio. Stochastic regression models generalize CAPM with the time dimension by allowing more factors than simply the return of market portfolio $R_m(t)$

$$R_i(t) = \alpha_i + \boldsymbol{\beta}_i' \boldsymbol{Z}(t) + \epsilon_i(t)$$

where $\boldsymbol{\beta}_i' = (\beta_{i,1}, \ldots, \beta_{i,p})'$ is a p-dimensional vector and $\{\boldsymbol{Z}(t) = (Z_1(t), \ldots, Z_p(t))'\}$ is factors process.

Empirical features of the realized returns for assets and market portfolios are well-known. SACF, which was defined in Chap. 1, of returns of IBM stock was plot in Fig. 1.1 (Sect. 1.1) and SACF of returns of S&P500 (squared transformed) is plot in Fig. 3.1.

From Figs. 1.1 and 3.1, we observe that the return of stock (i.e., IBM) shows the short memory dependence, and that squared transformed of a market index (i.e., S&P500) shows the long memory dependence.

Suppose that an q-dimensional process $\left\{\boldsymbol{Y}(t) = (Y_1(t), \ldots, Y_q(t))'\right\}$ is generated by

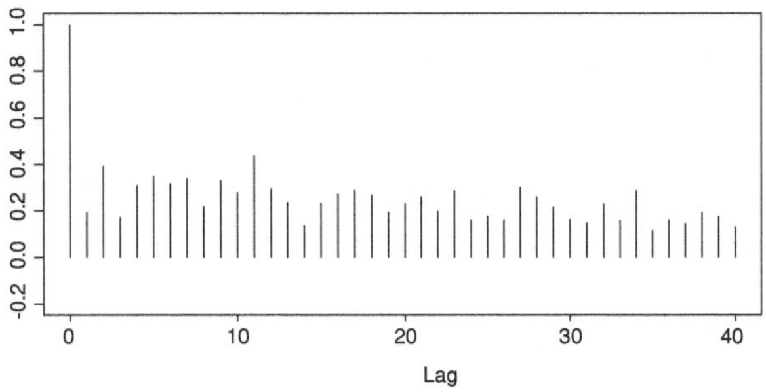

Fig. 3.1 SACF of return of S&P500 (square transformed). Taken from Amano et al. (2012). Published with the kind permission of © Tomoyuki Amano et al. (2012). Published under the creative commons attribution license

$$\boldsymbol{Y}(t) = \boldsymbol{\alpha} + \boldsymbol{B}'\boldsymbol{Z}(t) + \boldsymbol{\epsilon}(t), \quad (t = 1, 2, ..., n) \tag{3.36}$$

where $\boldsymbol{\alpha} = (\alpha_1, ..., \alpha_q)'$ and $\boldsymbol{B} = \{\beta_{ij}; i = 1, ..., p, j = 1, ..., q\}$ are unknown vector and matrix, respectively, $\{\boldsymbol{Z}(t) = (Z_1(t), ..., Z_p(t))'\}$ is a stochastic explanatory process and $\{\boldsymbol{\epsilon}(t) = (\epsilon_1(t), ..., \epsilon_q(t))'\}$ is a sequence of disturbance process. The ith component is written as

$$Y_i(t) = \alpha_i + \boldsymbol{\beta}_i'\boldsymbol{Z}(t) + \epsilon_i(t)$$

where $\boldsymbol{\beta}_i' = (\beta_{i,1}, ..., \beta_{i,p})$.

In the stochastic regression model, $\boldsymbol{Y}(t)$ is the return of assets, and $\boldsymbol{Z}(t)$ is the factors process. As we saw, empirical studies suggest that the return of assets $\{\boldsymbol{Y}(t)\}$ is short memory dependent and that the squared transformed return of the market portfolio is long memory dependent. Furthermore, as a market risk, the volatility of the return of the market portfolio should be considered and the stochastic regression model (3.36) in the case $\boldsymbol{Z}(t)$ is short memory dependent has been investigated enough. Then we assume that $\boldsymbol{Z}(t)$ is long memory dependent. On this ground, we investigate the conditions that the stochastic regression model (3.36) are well-defined. It is seen that if the model (3.36) is valid, we have to assume that $\{\boldsymbol{\epsilon}(t)\}$ is also long memory dependent and is correlated with $\{\boldsymbol{Z}(t)\}$.

Hence, we suppose that $\{\boldsymbol{Z}(t)\}$ and $\{\boldsymbol{\epsilon}(t)\}$ are defined by

$$\begin{aligned} \boldsymbol{Z}(t) &= \sum_{j=0}^{\infty} \boldsymbol{\gamma}(j)\boldsymbol{a}(t-j) + \sum_{j=0}^{\infty} \boldsymbol{\rho}(j)\boldsymbol{b}(t-j), \\ \boldsymbol{\epsilon}(t) &= \sum_{j=0}^{\infty} \boldsymbol{\eta}(j)\boldsymbol{e}(t-j) + \sum_{j=0}^{\infty} \boldsymbol{\xi}(j)\boldsymbol{b}(t-j), \end{aligned} \tag{3.37}$$

where $\{\boldsymbol{a}(t)\}$, $\{\boldsymbol{b}(t)\}$, and $\{\boldsymbol{e}(t)\}$ are p-dimensional zero-mean uncorrelated processes, and they are mutually independent. Here the coefficients $\{\boldsymbol{\gamma}(j)\}$ and $\{\boldsymbol{\rho}(j)\}$ are $p \times p$-matrices, and all the components of $\boldsymbol{\gamma}(j)$ are ℓ^1-summable, (for short, $\boldsymbol{\gamma}(j) \in \ell^1$), and those of $\boldsymbol{\rho}(j)$ are ℓ^2-summable (for short, $\boldsymbol{\rho}(j) \in \ell^2$). The coefficients $\{\boldsymbol{\eta}(j)\}$ and $\{\boldsymbol{\xi}(j)\}$ are $q \times p$-matrices, and $\boldsymbol{\eta}(j) \in \ell^1$ and $\boldsymbol{\xi}(j) \in \ell^2$. From (3.37) it follows that

$$Y(t) = \alpha + \sum_{j=0}^{\infty} \left(B'\gamma(j)a(t-j) + \eta(j)e(t-j)\right) + \sum_{j=0}^{\infty} \left(B'\rho(j) + \xi(j)\right) b(t-j).$$

Although $\left(B'\rho(j) + \xi(j)\right) \in \ell^2$ generally, if $B'\rho(j) + \xi(j) = O\left(\frac{1}{j^{\alpha}}\right)$, $\alpha > 1$, then $\left(B'\rho(j) + \xi(j)\right) \in \ell^1$, which leads to

Proposition 3.1 *If* $B'\rho(j)+\xi(j) = O\left(j^{-\alpha}\right)$, $\alpha > 1$, *then the process* $\{Y(t)\}$ *is short memory dependent.*

Proposition 3.1 provides an important view for the stochastic regression model, i.e., if we assume natural conditions on (3.36) based on the empirical studies, then they impose a sort of "curved structure": $B'\rho(j) + \xi(j) = O\left(j^{-\alpha}\right)$ on the regressor and disturbance. More important view is the statement implies that the process $\{\beta_i' Z(t) + \epsilon_i(t)\}$ is fractionally cointegrated. Here β_i and $\epsilon_i(t)$ are called the cointegrating vector and error, respectively (see Robinson and Yajima (2002)).

We discuss estimation of (3.36) satisfying Proposition 3.1. Since $E\left(Z(t)\epsilon(t)'\right) \neq \mathbf{0}$, the least squares estimator for B is known to be inconsistent. In what follows we assume that $\alpha = \mathbf{0}$ in (3.36), because it can be estimated consistently by the sample mean. However, by use of econometric theory, it is often possible to find other variables that are uncorrelated with the errors $\epsilon(t)$, which we call instrumental variables, and to overcome this difficulty. Without instrumental variables, correlations between the observables $\{Z(t)\}$ and unobservables $\{\epsilon(t)\}$ persistently contaminate our estimator for B. Hence, instrumental variables are useful in allowing us to estimate B.

Let $\{X(t)\}$ be r-dimensional vector $(p \leq r)$ instrumental variables with $E\{X(t)\} = \mathbf{0}$, $Cov\{X(t), Z(t)\} \neq \mathbf{0}$ and $Cov\{X(t), \epsilon(t)\} = \mathbf{0}$. Consider the *OLS* regression of $Z(t)$ on $X(t)$. If $Z(t)$ can be represented as

$$Z(t) = \delta' X(t) + u(t), \tag{3.38}$$

where δ is a $r \times p$ matrix and $\{u(t)\}$ is a p-dimensional vector process which is independent of $\{X(t)\}$, δ can be estimated by the OLS estimator

$$\hat{\delta} = \left[\sum_{t=1}^{n} X(t)X(t)'\right]^{-1} \left[\sum_{t=1}^{n} X(t)Z'(t)\right]. \tag{3.39}$$

From (3.36) with $\alpha = \mathbf{0}$ and (3.38), $Y(t)$ has the form

$$Y(t) = B'\delta' X(t) + B'u(t) + \epsilon(t)$$

and $\delta' X(t)$ is uncorrelated with $B'u(t) + \epsilon(t)$, hence B can be estimated by the OLS estimator

$$\hat{\boldsymbol{B}}_{OLS} = \left[\sum_{t=1}^{n} (\boldsymbol{\delta}'\boldsymbol{X}(t))\,(\boldsymbol{\delta}'\boldsymbol{X}(t))'\right]^{-1} \left[\sum_{t=1}^{n} (\boldsymbol{\delta}'\boldsymbol{X}(t))\,\boldsymbol{Y}'(t)\right]. \tag{3.40}$$

Using (3.39) and (3.40), we can propose the 2SLS estimator

$$\hat{\boldsymbol{B}}_{2SLS} = \left[\sum_{t=1}^{n} \left(\hat{\boldsymbol{\delta}}'\boldsymbol{X}(t)\right)\left(\hat{\boldsymbol{\delta}}'\boldsymbol{X}(t)\right)'\right]^{-1} \left[\sum_{t=1}^{n} \left(\hat{\boldsymbol{\delta}}'\boldsymbol{X}(t)\right)\boldsymbol{Y}'(t)\right]. \tag{3.41}$$

Now, we aim at proving the consistency and asymptotic normality of the 2SLS estimator $\hat{\boldsymbol{B}}_{2SLS}$. For this we assume that $\{\boldsymbol{\epsilon}(t)\}$ and $\{\boldsymbol{X}(t)\}$ jointly constitute the following linear process.

$$\begin{pmatrix} \boldsymbol{\epsilon}(t) \\ \boldsymbol{X}(t) \end{pmatrix} = \sum_{j=0}^{\infty} \boldsymbol{G}(j)\boldsymbol{\Gamma}(t-j) = \boldsymbol{A}(t) \ \ (say),$$

where $\{\boldsymbol{\Gamma}(t)\}$ is uncorrelated $(q+r)$-dimensional vector process with

$$E\{\boldsymbol{\Gamma}(t)\} = \boldsymbol{0}$$
$$E\{\boldsymbol{\Gamma}(t)\boldsymbol{\Gamma}(s)^*\} = \delta(t,s)\boldsymbol{K}$$
$$\delta(t,s) = \begin{cases} 1, & t = s \\ 0, & t \neq s \end{cases}$$

and $\boldsymbol{G}(j)$'s are $(q+r)\times(q+r)$ matrices which satisfy $\sum_{j=0}^{\infty} tr\,\{\boldsymbol{G}(j)\boldsymbol{K}\boldsymbol{G}(j)^*\} < \infty$. Then $\{\boldsymbol{A}(t)\}$ has the spectral density matrix

$$\boldsymbol{f}(\lambda) = \frac{1}{2\pi}\boldsymbol{k}(\lambda)\boldsymbol{K}\boldsymbol{k}(\lambda)^* = \{f_{ab}(\lambda);\, 1 \leq a, b \leq (q+r)\}, \quad (-\pi < \lambda \leq \pi),$$

where

$$\boldsymbol{k}(\lambda) = \sum_{j=0}^{\infty} \boldsymbol{G}(j)e^{i\lambda j} = \{k_{ab}(\lambda);\, 1 \leq a, b \leq (q+r)\}, \quad (-\pi < \lambda \leq \pi).$$

Further, we assume that $\int_{-\pi}^{\pi} \log\det \boldsymbol{f}(\lambda)d\lambda > -\infty$, so that the process $\{\boldsymbol{A}(t)\}$ is non-deterministic. For the asymptotics of $\hat{\boldsymbol{B}}_{2SLS}$, from page 108–109 of Hosoya (1997), we impose Assumption (HT) (ii), (iii), and (iv) in Chap. 1 on $\boldsymbol{\Gamma}(t)$ and the following Assumption 3.5.

Assumption 3.5 Each $f_{ab}(\lambda)$ is square-integrable.

Under above assumptions, we can establish the following theorem. For the proof, see Amano et al. (2012).

Theorem 3.3 *If* $\boldsymbol{\Gamma}(t)$ *satisfies Assumption (HT) (ii), (iii), and (iv) in Chap. 1, under Assumption 3.5 it holds that*

(i)

$$\hat{\boldsymbol{B}}_{2SLS} \xrightarrow{P} \boldsymbol{B},$$

(ii)

$$\sqrt{n}\left(\hat{\boldsymbol{B}}_{2SLS} - \boldsymbol{B}\right) \xrightarrow{d} \boldsymbol{Q}^{-1} E\left[\boldsymbol{Z}(t)\boldsymbol{X}'(t)\right] E\left[\boldsymbol{X}(t)\boldsymbol{X}(t)'\right]^{-1} \boldsymbol{U}$$

where

$$\boldsymbol{Q} = \left[E(\boldsymbol{Z}(t)\boldsymbol{X}(t)')\right]\left[E(\boldsymbol{X}(t)\boldsymbol{X}(t)')\right]^{-1}\left[E(\boldsymbol{X}(t)\boldsymbol{Z}(t)')\right]$$

and $\boldsymbol{U} = \{U_{i,j};\ 1 \le i \le r, 1 \le j \le q\}$ *is a random matrix whose elements follow normal distributions with mean* 0 *and*

$$\begin{aligned} Cov[U_{i,j}, U_{k,l}] = & 2\pi \int_{-\pi}^{\pi} [f_{q+i,q+k}(\lambda)\bar{f}_{j,l}(\lambda) + f_{q+i,l}(\lambda)\bar{f}_{j,q+k}(\lambda)]d\lambda \\ & + 2\pi \sum_{\beta_1,\ldots,\beta_4=1}^{p} \int_{-\pi}^{\pi}\int_{-\pi}^{\pi} \kappa_{q+i,\beta_1}(\lambda_1)\kappa_{j,\beta_2}(-\lambda_1) \\ & \times \kappa_{q+k,\beta_3}(\lambda_2)\kappa_{l,\beta_4}(-\lambda_2) Q^{\boldsymbol{\Gamma}}_{\beta_1,\ldots,\beta_4}(\lambda_1, -\lambda_2, \lambda_2) d\lambda_1 d\lambda_2, \end{aligned} \quad (3.42)$$

where $Q^{\boldsymbol{\Gamma}}_{\beta_1,\ldots,\beta_4}(\lambda_1, \lambda_2, \lambda_3)$ is the fourth order spectral density of $\boldsymbol{\Gamma}(t)$, which was defined in Chap. 1.

Next example prepares the asymptotic variance formula of $\hat{\boldsymbol{B}}_{2SLS}$ to investigate its features in simulation study.

Example 3.4 Let $\{Z(t)\}$ and $\{X(t)\}$ be scalar long memory processes, with spectral densities $\left\{2\pi|1 - e^{i\lambda}|^{2d_Z}\right\}^{-1}$ and $\left\{2\pi|1 - e^{i\lambda}|^{2d_X}\right\}^{-1}$, respectively, and cross spectral density $\frac{1}{2\pi}\left(1 - e^{i\lambda}\right)^{-d_X}\left(1 - e^{-i\lambda}\right)^{-d_Z}$, where $0 < d_Z < 1/2$ and $0 < d_X < 1/2$. Then

$$E\left(X(t)Z(t)\right) = \frac{1}{2\pi}\int_{-\pi}^{\pi} \frac{1}{\left(1 - e^{i\lambda}\right)^{d_X}} \frac{1}{\left(1 - e^{-i\lambda}\right)^{d_Z}} d\lambda.$$

Suppose that $\{\epsilon(t)\}$ is a scalar uncorrelated process with $\sigma_\epsilon^2 \equiv E\left\{\epsilon(t)^2\right\}$. Assuming Gaussianity of $\{\boldsymbol{A}(t)\}$, it is seen that the right hand of (3.42) is

$$2\pi \int_{-\pi}^{\pi} \frac{1}{2\pi|1 - e^{i\lambda}|^{2d_X}} \frac{\sigma_\epsilon^2}{2\pi} d\lambda,$$

which entails

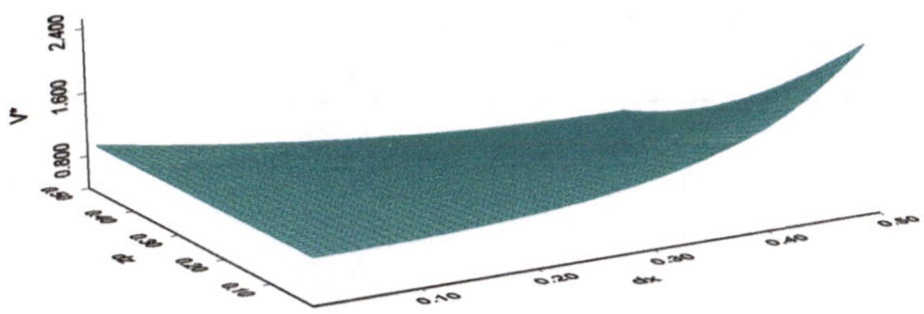

Fig. 3.2 $V_*(d_x, d_z)$ in Example 3.5. Taken from Amano et al. (2012). Published with the kind permission of © Tomoyuki Amano et al. 2012. Published under the creative commons attribution license

$$\lim_{n\to\infty} Var\left\{\sqrt{n}\left(\hat{B}_{2SLS} - B\right)\right\} = \frac{2\pi \int_{-\pi}^{\pi} \frac{1}{2\pi|1-e^{i\lambda}|^{2d_X}} \frac{\sigma_\epsilon^2}{2\pi} d\lambda}{\left(\frac{1}{2\pi}\int_{-\pi}^{\pi} \frac{1}{(1-e^{i\lambda})^{d_X}} \frac{1}{(1-e^{-i\lambda})^{d_Z}} d\lambda\right)^2}$$

$$= \sigma_\epsilon^2 \left(2\pi \frac{\int_{-\pi}^{\pi} \frac{1}{|1-e^{i\lambda}|^{2d_X}} d\lambda}{\left(\int_{-\pi}^{\pi} \frac{1}{(1-e^{i\lambda})^{d_X}} \frac{1}{(1-e^{-i\lambda})^{d_Z}} d\lambda\right)^2} \right)$$

$$= \sigma_\epsilon^2 \times V_*\left(d_X, d_Z\right).$$

Now, we evaluate the behavior of $\hat{\boldsymbol{B}}_{2SLS}$ in the case $p = 1$ in (3.36) numerically.

Example 3.5 Under the condition of Example 3.4, we investigate the asymptotic variance behavior of $\hat{B}_{2SLS}$ by simulation. Figure 3.2 plots $V_*\left(d_X, d_Z\right)$ for $0 < d_X < \frac{1}{2}$ and $0 < d_Z < \frac{1}{2}$.

From Fig. 3.2, we observe that if $d_Z \searrow 0$ and if $d_X \nearrow 1/2$, then V_* becomes large and otherwise V_* is small. This result implies only in the case that the long memory behavior of $Z(t)$ is weak and the long memory behavior of $X(t)$ is strong, V_* is large. Note that long memory behavior of $Z(t)$ makes the asymptotic variance of the 2SLS estimator small but one of $X(t)$ makes it large.

Example 3.6 We consider the following model,

$$Y(t) = Z(t) + \epsilon(t),$$
$$Z(t) = X(t) + u(t),$$
$$\epsilon(t) = w(t) + u(t),$$

where $X(t)$, $w(t)$, and $u(t)$ are scalar long memory processes which follow FARIMA(0, d_1, 0), FARIMA(0, d_2, 0), and FARIMA(0, 0.1, 0), respectively. Note that $Z(t)$ and $\epsilon(t)$ are correlated, $X(t)$ and $Z(t)$ are correlated, but $X(t)$ and $\epsilon(t)$ are independent. Under this model, we compare $\hat{B}_{2SLS}$ with the ordinary least squares estimator $\tilde{B}_{OLS}$ for B, which is defined as

$$\tilde{B}_{OLS} = \left[\sum_{t=1}^{n} Z^2(t)\right]^{-1} \left[\sum_{t=1}^{n} Z(t)Y(t)\right].$$

Table 3.5 MSE of $\hat{B}_{2SLS}$ and $\tilde{B}_{OLS}$

d_2	0.1	0.2	0.3
$\hat{B}_{2SLS}$ $(d_1 = 0.1)$	0.03	0.052	0.189
$\tilde{B}_{OLS}$ $(d_1 = 0.1)$	0.259	0.271	0.34
$\hat{B}_{2SLS}$ $(d_1 = 0.2)$	0.03	0.075	0.342
$\tilde{B}_{OLS}$ $(d_1 = 0.2)$	0.178	0.193	0.307
$\hat{B}_{2SLS}$ $(d_1 = 0.3)$	0.019	0.052	0.267
$\tilde{B}_{OLS}$ $(d_1 = 0.3)$	0.069	0.089	0.23

Taken from Amano et al. (2012). Published with the kind permission of © Tomoyuki Amano et al. 2012. Published under the creative commons attribution license

Table 3.6 $\hat{\boldsymbol{B}}_{2SLS}$ based on the actual financial data

Stock	IBM	Nike	Amazon	American Express	Ford
$\hat{\boldsymbol{B}}_{2SLS}$	0.75	1.39	1.71	2.61	−1.89

Taken from Amano et al. (2012). Published with the kind permission of © Tomoyuki Amano et al. 2012. Published under the creative commons attribution license

The length of $X(t)$, $Y(t)$, and $Z(t)$ are set by 100, and based on 5,000 times simulation we report MSE of $\hat{B}_{2SLS}$ and $\tilde{B}_{OLS}$. We set $d_1, d_2 = 0.1, 0.2, 0.3$ in Table 3.5.

In most cases of d_1 and d_2 in Table 3.5, MSE of $\hat{B}_{2SLS}$ is smaller than that of $\tilde{B}_{OLS}$. Hence from this Example we can see our estimator $\hat{B}_{2SLS}$ is better than $\tilde{B}_{OLS}$ in the sense of MSE. Furthermore from Table 3.5, we can see that MSE of $\hat{B}_{2SLS}$ and $\tilde{B}_{OLS}$ increases as d_2 becomes large, that is, long memory behavior of $w(t)$ makes the asymptotic variances of $\hat{B}_{2SLS}$ and $\tilde{B}_{OLS}$ large.

Example 3.7 In this example, we calculate $\hat{\boldsymbol{B}}_{2SLS}$ based on the actual financial data. We choose S&P500 (square transformed) as $Z(t)$ and Nikkei Average as an instrumental variable $X(t)$. Assuming that $\boldsymbol{Y}(t)$ (5×1) consists of the return of IBM, Nike, Amazon, American Expresses, and Ford, the 2SLS estimates for B_i, $i = 1, \ldots, 5$ are recorded in Table 3.6. We chose Nikkei Stock Average as the instrumental variable, because we got the following correlation analysis between the residual processes of returns and Nikkei:
Correlation of IBM's residual and Nikkei Average's return: −0.000311
Correlation of Nike's residual and Nikkei Average's return: −0.00015
Correlation of Amazon's residual and Nikkei Average's return: −0.000622
Correlation of American Express's residual and Nikkei Average's return: 0.000147
Correlation of Ford's residual and Nikkei Average's return: −0.000536,
which supports the assumption $Cov(X(t), \boldsymbol{\epsilon}(t)) = \boldsymbol{0}$.

From Table 3.6, we observe that the return of American Express is strongly correlated with that of S&P500 and the return of the auto industry stock (Ford) is negatively correlated with that of S&P500.

References

Amano, T., Kato, T., Taniguchi, M.: Statistical estimation for CAPM with long-memory dependence. Adv. Decis. Sci. **2012**, Article ID 571034 (2012)

Amano, T., Taniguchi, M.: Control variate method for stationary processes. J. Econometrics **165**, 20–29 (2011)

Black, F.: Capital market equilibrium with restricted borrowing. J. Bus. **45**, 444–455 (1972)

Brillinger, D.R.: Time Series: Data Analysis and Theory, Expanded edn. Holden-Day, San Francisco (2001)

Campbell, J.Y., Lo, A.W., Mackinlay, A.C.: The Econometrics of Financial Markets. Princeton University Press, New Jersey (1997)

Chan, N.H., Wong, H.Y.: Simulation Techniques in Financial Risk Management. Wiley, New York (2006)

Choy, K., Taniguchi, M.: Stochastic regression model with dependent disturbances. J. Time Ser. Anal. **22**, 175–196 (2001)

Geary, R.C.: Determination of linear relations between systematic parts of variables with errors of observation the variances of which are unknown. Econometrica **17**, 30–58 (1949)

Glasserman, P.: Monte Carlo Methods in Financial Engineering. Springer, New York (2004)

Hosoya, Y.: A limit theory for long-range dependence and statistical inference on related models. Ann. Stat. **25**, 105–137 (1997)

Lavenberg, S.S., Welch, P.D.: A perspective on the use of control variables to increase the efficiency of Monte Carlo simulations. Manag. Sci. **27**, 322–335 (1981)

Lintner, J.: The valuation of risk assets and the selection of risky investments in stock portfolios and capital budgets. Rev. Econ. Stat. **47**, 13–37 (1965)

Markowitz, H.: Portfolio Selection: Efficient Diversification of Investments. Wiley, New York (1991)

Nelson, B.L.: Control variate remedies. Oper. Res. **38**, 974–992 (1990)

Reiersöl, O.: Confluence analysis by means of instrumental sets of variables. Akiv för Matematik, Astronomi och Fysik. **32A**, 1–119 (1945)

Robinson, P.M., Hidalgo, F.J.: Time series regression with long-range dependence. Ann. Stat. **25**, 77–104 (1997)

Robinson, P.M., Yajima, Y.: Determination of cointegrating rank in fractional system. J. Econometrics **106**, 217–241 (2002)

Rubinstein, R.Y., Marcus, R.: Efficiency of multivariate control variates in Monte Carlo simulation. Oper. Res. **33**, 661–677 (1985)

Sharpe, W.: Capital asset prices: a theory of market equilibrium under conditions of risk. J. Finance **19**, 425–442 (1964)

White, H.: Asymptotic Theory for Econometricians. Academic Press, New York (2001)

Wright, P.G.: The Tariff on Animal and Vegetable Oils. MacMillan, New York (1928)

References

[illegible]

Chapter 4
Some Techniques for ARCH Financial Time Series

Abstract This chapter introduces two techniques which can be utilized in study of financial risks. The first one is the method called Quantile Regression (QR), which can be used to analyze the conditional quantile of financial assets. There, by means of rank-based semiparametrics, we provide the statistically efficient inference under the autoregressive conditional heteroskedasticity (ARCH). The second technique, the realized volatility, estimates the conditional variance, or "volatility" of financial assets. Revealing the fact that its inference can be greatly affected by the existence of additional noize called market microstructure, we introduce and study the asymptotics of some appropriate estimator under the microstructure with ARCH-dependent structure.

Keywords ARCH model · Quantile regression · Rank-based semiparametrics · Realized volatility · Market microstructure

4.1 Introduction

In this chapter, two techniques for analysis of financial risks will be introduced. For both techniques, namely, the Quantile Regression (QR) with its semiparametrics and Realized Volatility (RV) with dependent microstructure, here we extend their applicability to the AutoRegressive Conditional Heteroskedasticity (ARCH) models. As is well known, ARCH structure is one of the most particular characteristics of financial asset's prices. (See, e.g., Gouriéroux (1997).)

The construction of this chapter is as follows. In Sect. 4.2, we explain how QR method can be used for ARCH time series and provide the semiparametrically efficient inference as well. This method allows us to estimate the conditional quantiles of financial assets, such as some variants of Value-at-Risk. Section 4.3 is concerned with estimation of the conditional variance, which is called "volatility" in finance. There, the existence of additional noize called market microstructure affects our inference greatly. By introducing some appropriate estimator, we show that this estimator works well in the case where microstructure exhibits the ARCH property.

M. Taniguchi et al., *Statistical Inference for Financial Engineering*, SpringerBriefs in Statistics, DOI: 10.1007/978-3-319-03497-3_4, © The Author(s) 2014

4.2 Quantile Regression and Its Semiparametric Efficiency for ARCH Series

Motivated by squared ARCH models, we investigate a Quantile AutoRegression (QAR) model for non-negative series (Koenker and Bassett (1978); Koenker and Xiao (2006); Koenker (2005)). This QAR can be interpreted as a Random Coefficient AutoRegression (RCAR) model for which we see the Local Asymptotic Normality (LAN). More precisely, we decompose the random coefficient as a product of two components, a quantile coefficient $\boldsymbol{\gamma}(\tau)$ for fixed τ and a "standardized random coefficient" with τ-quantile one, and think of the former as a parameter of interest while the density g_1 of the latter is treated as a nuisance. Having the LAN result with respect to $\boldsymbol{\gamma}(\tau)$ for each τ-th quantile restriction (identification constraint) given on the nuisance, we follow the one-step estimation method of Le Cam. Consequently, we provide a semiparametrically efficient (at g_1) version of QAR estimators.

4.2.1 Model, Estimators, and Some Asymptotics

Suppose that we have observation $\{y_t^2\}_{t=1}^n$ from some experiment $(\mathbb{R}_+^n, \mathscr{B}^{(n)}, \mathcal{P}^{(n)})$. Also, let us consider a sequence of probability measures $\mathcal{P}^{(n)} = \{P_{\boldsymbol{\beta},g}^{(n)} | \boldsymbol{\beta} \in \boldsymbol{\Theta}, g \in \mathcal{G}\}$ which is described by the model known as ARCH(1), that is,

$$Y_t^2 = (\beta_0 + \beta_1 Y_{t-1}^2) Z_t^2, \tag{4.1}$$

where $\boldsymbol{\beta} := (\beta_0, \beta_1)'$, and $\{Z_t^2\}_{t=1}^n$ are i.i.d. with density g and distribution function G. Here $\boldsymbol{\Theta} \subset \mathbb{R}^2$ and $\mathcal{G}$ are some parameter space and family of probability density functions, respectively. Note that we introduced the innovation density g as that of the squared process $\{Z_t^2\}_{t=1}^n$, not that of $\{Z_t\}_{t=1}^n$ itself. The conditional τ-quantile of Y_t^2 (conditioned by the realized value y_{t-1}^2) in model (4.1) is

$$F_{Y_t^2}^{-1}(\tau | y_{t-1}^2) := F_{Y_t^2 | Y_{t-1}^2 = y_{t-1}^2}^{-1}(\tau) = (\beta_0 + \beta_1 y_{t-1}^2) G^{-1}(\tau),$$

where $F_{X|S}^{-1}(\cdot)$ denotes the (conditional) quantile function, i.e., $F_{X|S}^{-1}(\tau) := \inf\{x : P(X \le x|S) \ge \tau\}$.

Motivated by the squared series of ARCH(1) above, let $X_t \ge 0, t \in \mathbb{Z}$, be a process which takes non-negative values. Then we define a Quantile AutoRegression (QAR) model, for given $\tau \in (0, 1)$, by

$$F_{X_t}^{-1}(\tau | x_{t-1}) = \gamma_0(\tau) + \gamma_1(\tau) x_{t-1}, \tag{4.2}$$

where x_{t-1} is the realized value of X_{t-1} and $\boldsymbol{\gamma}(\tau) := (\gamma_0(\tau), \gamma_1(\tau))'$ is the parameter of interest. So, we see that the model (4.1) is a special case of (4.2) whith

$\boldsymbol{\gamma}(\tau) \equiv \boldsymbol{\beta} G^{-1}(\tau)$. In the following, although we discuss this quantile regression model of order 1 for simplicity, our method can be easily extended to the case of higher dependency so that it can be applicable to ARCH(p), $p \geq 1$.

Now, we define the QR estimator $\hat{\boldsymbol{\gamma}}_n(\tau)$ as

$$\hat{\boldsymbol{\gamma}}_n(\tau) := \underset{b_0, b_1}{\operatorname{argmin}} \sum_{t=1}^{n} \rho_\tau \left(X_t - (b_0 + b_1 X_{t-1}) \right), \tag{4.3}$$

where $\rho_\tau(u) := \tau |u| \cdot 1\{u \geq 0\} + (1-\tau)|u| \cdot 1\{u < 0\} = u(\tau - 1\{u < 0\})$, $u \in \mathbb{R}$ is called the *check function* (see Koenker (2005)), and $1A$ is the indicator function defined by $1A = 1_A(\omega) := 1$ if $\omega \in A$, $:= 0$ if $\omega \notin A$. Also, defining

$$D_0 := E\left[\boldsymbol{X}_{t-1} \boldsymbol{X}_{t-1}'\right], \quad \text{and}$$
$$D_{\boldsymbol{\gamma}(\tau)} := E\left[f_{X_t}(\boldsymbol{X}_{t-1}' \boldsymbol{\gamma}(\tau) | \boldsymbol{X}_{t-1}) \boldsymbol{X}_{t-1} \boldsymbol{X}_{t-1}'\right]$$

with $\boldsymbol{X}_{t-1} := (1, X_{t-1})'$, some regularity conditions are now in order:

Assumption 4.1

(i) There exists a unique strictly stationary solution $\{X_t\}$ of model (4.2) with $E[X_t^2] < \infty$;
(ii) The matrices D_0 and $D_{\boldsymbol{\gamma}(\tau)}$ are nonsingular;
(iii) For $\tau = \tau_0$ of our interest, the density of X_t conditional on $\boldsymbol{X}_{t-1}$, denoted by $f_{X_t}(x|\boldsymbol{X}_{t-1})$, is continuous and positive at the points $x = F_{X_t}^{-1}(\tau_0 | \boldsymbol{X}_{t-1})$, $t = 1, 2, \ldots$.

Proposition 4.1 *Let Assumption 4.1 hold. Then, for any* $\tau \in (0, 1)$, *as* $n \to \infty$,

$$\sqrt{n}(\hat{\boldsymbol{\gamma}}_n(\tau) - \boldsymbol{\gamma}(\tau)) \xrightarrow{d} N\Big(\boldsymbol{0}, \ \tau(1-\tau) D_{\boldsymbol{\gamma}(\tau)}^{-1} D_0 D_{\boldsymbol{\gamma}(\tau)}^{-1}\Big), \tag{4.4}$$

Proof See e.g., Koenker (2005, Chap. 4), and Koul (1992, Eq. (7.3b.6)). □

Now, observe that, for each fixed τ, the QR coefficient $\boldsymbol{\gamma}(\tau)$ can be characterized as the parameter $\boldsymbol{\gamma} = (\gamma_0, \gamma_1)'$ of some model such as

$$X_t = (\gamma_0 + \gamma_1 X_{t-1}) \xi_t, \tag{4.5}$$

where $\{\xi_t \geq 0, t \in \mathbb{Z}\}$ is i.i.d. with distribution function G_1 and density g_1,

$$g_1 \in \mathcal{F}^\tau := \left\{ f : [0, \infty) \to [0, \infty) \,\middle|\, \int_0^1 f(x)dx = \tau = 1 - \int_1^\infty f(x)dx \right\}.$$

This "quantile-restricted ARCH model" is a fixed-τ submodel of general QAR model (4.2). There we find that

$$f_{X_t}(\boldsymbol{X}'_{t-1}\boldsymbol{\gamma}(\tau)|\boldsymbol{X}_{t-1}) = \frac{f_{X_t/\boldsymbol{X}'_{t-1}\boldsymbol{\gamma}}(1)}{\boldsymbol{X}'_{t-1}\boldsymbol{\gamma}} \left(= \frac{g_1(G_1^{-1}(\tau))}{\boldsymbol{X}'_{t-1}\boldsymbol{\gamma}} \right),$$

so that the following corollary holds from Proposition 4.1.

Corollary 4.1 *Let Assumption 4.1 hold. Then, for any $\boldsymbol{\gamma} \in \boldsymbol{\Theta} \subset \mathbb{R}_+^2$ and any $g_1 \in \mathcal{F}^\tau$, as $n \to \infty$,*

$$\sqrt{n}(\hat{\boldsymbol{\gamma}}_n(\tau) - \boldsymbol{\gamma}) \xrightarrow{d} N\left(\mathbf{0}, \ \frac{\tau(1-\tau)}{g_1^2(1)} D_{1,\boldsymbol{\gamma}}^{-1} D_0 D_{1,\boldsymbol{\gamma}}^{-1}\right), \tag{4.6}$$

where $D_{1,\boldsymbol{\gamma}} := E[\boldsymbol{X}_{t-1}\boldsymbol{X}'_{t-1}/\boldsymbol{X}'_{t-1}\boldsymbol{\gamma}]$.

Remark 4.1 Note here that, for the submodel (4.5), Assumption 4.1-(i) is satisfied if the following conditions are fulfilled

(i) The "true" values γ_0 and γ_1 are positive;
(ii) $E[\xi_0] < \infty$, $E[\xi_0^2] < \infty$ and $\left(E[\xi_0^2]\right)^{1/2}\gamma_1 < 1$.

The latter is a sufficient condition due to Giraitis et al. (2000), which may be easy to verify. Necessary and sufficient conditions can also be found in He and Teräsvirta (1999). ■

Now, as a preparation for Sect. 4.2.2, let us write $\mathcal{P}_{\boldsymbol{\gamma}}^{(n)} := \{P_{\boldsymbol{\gamma},g_1}^{(n)} | g_1 \in \mathcal{F}^\tau\}$ for the fixed-$\boldsymbol{\gamma}$ subfamily of $\mathcal{P}^{(n)}$. In order to apply the general results of Hallin et al. (2006b), we first establish the existence of a generating group of $\mathcal{P}_{\boldsymbol{\gamma}}^{(n)}$. Denoting by $\mathcal{H}_1$ the set of all continuous and strictly increasing functions h from $\mathbb{R}_+$ to $\mathbb{R}_+$ satisfying $h(0) = 0$ and $h(1) = 1$, define

$$a_h^{(n)}(\xi_1, \ldots, \xi_n) := (h(\xi_1), \ldots, h(\xi_n))$$

and consider the transformation group $A_{\boldsymbol{\gamma}}^{(n)}$ (acting on $\mathbb{R}_+^n$)

$$A_{\boldsymbol{\gamma}}^{(n)} := \left\{ (\boldsymbol{\xi}_{\boldsymbol{\gamma}}^{(n)})^{-1} \circ a_h^{(n)} \circ \boldsymbol{\xi}_{\boldsymbol{\gamma}}^{(n)}, h \in \mathcal{H}_1 \right\}, \tag{4.7}$$

where $\boldsymbol{\xi}_{\boldsymbol{\gamma}}^{(n)}$ is a residual function defined by

$$\boldsymbol{\xi}_{\boldsymbol{\gamma}}^{(n)}(X_1, \ldots, X_n) := \left(\frac{X_1}{\gamma_0 + \gamma_1 X_0}, \ldots, \frac{X_n}{\gamma_0 + \gamma_1 X_{n-1}} \right) = (\xi_{\boldsymbol{\gamma},1}, \ldots, \xi_{\boldsymbol{\gamma},n}).$$

Then, $A_{\boldsymbol{\gamma}}^{(n)}$ is a generating group for $\mathcal{P}_{\boldsymbol{\gamma}}^{(n)}$, in the sense that for all $P_1, P_2 \in \mathcal{P}_{\boldsymbol{\gamma}}^{(n)}$, there exists $a \in A_{\boldsymbol{\gamma}}^{(n)}$ such that $(X_1, \ldots, X_n) \overset{d}{\sim} P_1$ iff $a(X_1, \ldots, X_n) \overset{d}{\sim} P_2$. So, defining $S_{\boldsymbol{\gamma},t}$ and $R_{\boldsymbol{\gamma},t}^{(n)}$ as the sign and the rank (among n copies) of $(\xi_{\boldsymbol{\gamma},t} - 1)$ respectively, we see that the map

$$T(X_1, \ldots, X_n) = (S_{\gamma,1}, \ldots, S_{\gamma,n}; R^{(n)}_{\gamma,1}, \ldots, R^{(n)}_{\gamma,n}), \tag{4.8}$$

is maximal invariant w.r.t. $A^{(n)}_{\gamma}$ (cf. Schmetterer (1974, Sections 7.4 and 7.6)). Maximal invariance implies that, for any invariant map T' (i.e., $T'(\boldsymbol{x}) = T'(a\boldsymbol{x})$, $a \in A^{(n)}_{\gamma}$), there exists a measurable function χ such that $T' = \chi \circ T$. Consequently, we have that any test which is invariant w.r.t. $A^{(n)}_{\gamma}$ is $\mathcal{B}^{(n)}_{\gamma}$-measurable, where

$$\mathcal{B}^{(n)}_{\gamma} := \sigma(S_{\gamma,1}, \ldots, S_{\gamma,n}; R^{(n)}_{\gamma,1}, \ldots, R^{(n)}_{\gamma,n}) \tag{4.9}$$

which is the σ-field generated by $\{S_{\gamma,t}\}_{t=1}^{n}$ and $\{R^{(n)}_{\gamma,t}\}_{t=1}^{n}$. This $\mathcal{B}^{(n)}_{\gamma}$ will play a key role in Sect. 4.2.2.

4.2.2 Semiparametrically Efficient Inference

In this section, we improve the efficiency of $\hat{\boldsymbol{\gamma}}_n(\tau)$ with regard to the quantile-restricted ARCH model (4.5). Since here we do not have any knowledge about the true density g_1, except for the fact that it belongs to $\mathcal{F}^{\tau}$, we arbitrarily choose a "reference density" f from $\mathcal{F}^{\tau}$, and correspondingly define a "reference model"

$$X_t = (\gamma_0 + \gamma_1 X_{t-1})\xi_t, \tag{4.10}$$

where $X_t \geq 0$, $t \in \mathbb{Z}$, and $\{\xi_t \geq 0, t \in \mathbb{Z}\}$ is i.i.d. with density $f \in \mathcal{F}^{\tau}$. The goal is to construct an asymptotically efficient version of $\hat{\boldsymbol{\gamma}}_n(\tau)$ based on some feasible $f \in \mathcal{F}^{\tau}$, that is, attaining the semi-parametric lower bound at correctly specified density $f = g_1$ that nevertheless remains $\sqrt{n}$-consistent under misspecified density ($f \neq g_1$).

First, let us assume stationarity by assuming, e.g., that the conditions of Remark 4.1 hold. Also, in order to ensure the regularity of the reference model (4.10), we require the following assumption, which is essentially same as that for standard (i.e., nonsquared) ARCH models since $\boldsymbol{\gamma}$ has no effect on the sign of Y_t.

Assumption 4.2

(i) The distribution of the initial value, $\mathscr{L}(X_0|\cdot)$, say, is continuous in probability with respect to $\boldsymbol{\gamma}$, i.e., for any $\boldsymbol{\gamma}_n \to \boldsymbol{\gamma}$,

$$\mathscr{L}(X_0|\boldsymbol{\gamma}_n) \xrightarrow{P} \mathscr{L}(X_0|\boldsymbol{\gamma});$$

(ii) the density $f \in \mathcal{F}^{\tau}$ is absolutely continuous with derivative f' and has finite Fisher information for scale, i.e.,

$$0 < \mathcal{I}_f := \int_0^\infty \varphi_f^2(x) f(x) dx = \int_0^1 \varphi_f^2(F^{-1}(u)) du < \infty,$$
$$\text{where} \quad \varphi_f(x) := -1 - \frac{x f'(x)}{f(x)}.$$

Then, by Drost et al. (1997, Theorem 2.1), the model (4.10) satisfies the uniform LAN (ULAN) condition under $f \in \mathcal{F}^\tau$ and for any $\boldsymbol{\gamma}_n$ of the form $\boldsymbol{\gamma} + O(n^{-1/2})$, with the central sequence

$$\boldsymbol{\Delta}^{(n)}_{\boldsymbol{\gamma}_n, f} := \frac{1}{\sqrt{n}} \sum_{t=1}^{n} \varphi_f(\xi_{\boldsymbol{\gamma}_n,t}) \boldsymbol{W}_t(\boldsymbol{\gamma}_n), \quad \boldsymbol{W}_t(\boldsymbol{\gamma}_n) \equiv \frac{\boldsymbol{X}_{t-1}}{\boldsymbol{X}'_{t-1}\boldsymbol{\gamma}_n}, \tag{4.11}$$

and Fisher information $\mathcal{I}_f(\boldsymbol{\gamma}) \equiv \mathcal{I}_f \cdot E[\boldsymbol{W}_t(\boldsymbol{\gamma})\boldsymbol{W}'_t(\boldsymbol{\gamma})]$, where $\xi_{\boldsymbol{\gamma}_n,t} := X_t/(\gamma_{n,0} + \gamma_{n,1} X_{t-1})$ denotes the residuals. The reason why we have stated ULAN, rather than LAN at single $\boldsymbol{\gamma}$, is due to the one-step improvement, which will be discussed below.

Here we observe that, by recursively replacing X_{t-i} by $(\gamma_0 + \gamma_1 X_{t-i-1})\xi_{\boldsymbol{\gamma},t-i}$, recovering the exact X_{t-i} requires observing the infinitely many lagged residuals $\{\xi_{\boldsymbol{\gamma},t-i}\}_{i=1}^\infty$. However, this can be overcome by re-expressing (up to $o_P(1)$ terms) the central sequence (4.11) by another central sequence which involves finitely lagged residuals as follows. Since $X_{t-1} = \gamma_0 \sum_{i=1}^{t-1} \gamma_1^{i-1} \prod_{j=1}^{i} \xi_{\boldsymbol{\gamma},t-j}$, we have for the second component (first component can be treated similarly) that

$$E\left| \frac{X_{t-1}}{\gamma_0 + \gamma_1 X_{t-1}} - \frac{\gamma_0 \sum_{i=1}^{p} \gamma_1^{i-1} \prod_{j=1}^{i} \xi_{\boldsymbol{\gamma},t-j}}{\gamma_0 + \gamma_1 \gamma_0 \sum_{i=1}^{p} \gamma_1^{i-1} \prod_{j=1}^{i} \xi_{\boldsymbol{\gamma},t-j}} \right|$$
$$\leq E\left[\sum_{i=p+1}^{t-1} \gamma_1^{i-1} \prod_{j=1}^{i} \xi_{\boldsymbol{\gamma},t-j} \right] \leq \frac{\sigma_Z^2}{1 - \gamma_1 \sigma_Z^2} \left(\gamma_1 \sigma_Z^2\right)^p, \tag{4.12}$$

where $\sigma_Z^2 := E[\xi_{\boldsymbol{\gamma},t}]$. By the stationarity, we know that (4.12) converges to zero at a geometric rate as $p = p(n) \to \infty$, where this convergence $p \to \infty$ can be made arbitrarily slow. Hence, (4.12) is $o_P(1)$ as $n \to \infty$ and so we do not lose anything through this re-expression.

Then, following Hallin and Werker (2003), a semi-parametrically efficient procedure can be obtained by projecting (4.11) on the maximal invariant σ-field, which is $\mathcal{B}^{(n)}_{\boldsymbol{\gamma}}$ in our case. So, together with (4.12), "good" inference should be based on

$$\underset{\sim}{\boldsymbol{\Delta}}^{(n)}_{\boldsymbol{\gamma},f} := E^{(n)}_{\boldsymbol{\gamma},f}[\boldsymbol{\Delta}^{(n)}_{\boldsymbol{\gamma},f} | \mathcal{B}^{(n)}_{\boldsymbol{\gamma}}]$$
$$= \frac{1}{\sqrt{n-p}} \sum_{t=p+1}^{n} E^{(n)}_{\boldsymbol{\gamma},f}\left[\varphi_f(\xi_{\boldsymbol{\gamma},t}) \frac{[1, \gamma_0 \sum_{i=1}^{p} \gamma_1^{i-1} \prod_{j=1}^{i} \xi_{\boldsymbol{\gamma},t-j}]'}{\gamma_0 + \gamma_1\gamma_0 \sum_{i=1}^{p} \gamma_1^{i-1} \prod_{j=1}^{i} \xi_{\boldsymbol{\gamma},t-j}} \Big| \mathcal{B}^{(n)}_{\boldsymbol{\gamma}} \right] + o_P(1)$$
$$= \frac{1}{\sqrt{n-p}} \sum_{t=p+1}^{n} E^{(n)}_{\boldsymbol{\gamma},f}\Big[\varphi_f[F^{-1}(U_{\boldsymbol{\gamma},t})]$$

$$\times \frac{[1, \gamma_0 \sum_{i=1}^{p} \gamma_1^{i-1} \prod_{j=1}^{i} F^{-1}(U_{\boldsymbol{\gamma},t-j})]'}{\gamma_0 + \gamma_1\gamma_0 \sum_{i=1}^{p} \gamma_1^{i-1} \prod_{j=1}^{i} F^{-1}(U_{\boldsymbol{\gamma},t-j})} |\mathcal{B}_{\boldsymbol{\gamma}}^{(n)}\Bigg] + o_P(1)$$

$$\stackrel{(say)}{=} \frac{1}{\sqrt{n-p}} \sum_{t=p+1}^{n} E_{\boldsymbol{\gamma},f}^{(n)}[\boldsymbol{\psi}_{f,p}(U_{\boldsymbol{\gamma},t}, \ldots, U_{\boldsymbol{\gamma},t-p})|\mathcal{B}_{\boldsymbol{\gamma}}^{(n)}] + o_P(1), \tag{4.13}$$

where $U_{\boldsymbol{\gamma},t} := F(\xi_{\boldsymbol{\gamma},t})$ which is i.i.d. uniform on [0, 1] under $P_{\boldsymbol{\gamma},f}^{(n)}$. The quantity (4.13), which involves the expectation of $\boldsymbol{\psi}_{f,p}$, is said to be made of "exact scores". A more convenient version, i.e., those made of "approximate scores", is

$$\frac{1}{\sqrt{n-p}} \sum_{t=p+1}^{n} \boldsymbol{\psi}_{f,p}(V_{\boldsymbol{\gamma},t}^{(n)}, \ldots, V_{\boldsymbol{\gamma},t-p}^{(n)}) + o_P(1), \tag{4.14}$$

where

$$V_{\boldsymbol{\gamma},t}^{(n)} := \begin{cases} \tau \cdot \dfrac{R_{\boldsymbol{\gamma},t}^{(n)}}{N_{\boldsymbol{\gamma},L}^{(n)} + 1} & \text{if } R_{\boldsymbol{\gamma},t}^{(n)} \le N_{\boldsymbol{\gamma},L}^{(n)}, \\ \tau + (1-\tau) \cdot \dfrac{R_{\boldsymbol{\gamma},t}^{(n)} - N_{\boldsymbol{\gamma},L}^{(n)}}{n - N_{\boldsymbol{\gamma},L}^{(n)} + 1} & \text{otherwise,} \end{cases}$$

with $N_{\boldsymbol{\gamma},L}^{(n)} := \#\{t \in \{1, \ldots, n\} | S_{\boldsymbol{\gamma},t} = -1\}$. In short, we are first rewriting the residual $\xi_{\boldsymbol{\gamma},t}$ as $F^{-1}(U_{\boldsymbol{\gamma},t})$ with realization $U_{\boldsymbol{\gamma},t}$ of [0, 1]-uniform r.v., then approximating those $U_{\boldsymbol{\gamma},t}$ by $V_{\boldsymbol{\gamma},t}^{(n)}$ with $\{N_{\boldsymbol{\gamma},L}^{(n)}; R_{\boldsymbol{\gamma},1}^{(n)}, \ldots, R_{\boldsymbol{\gamma},n}^{(n)}\}$ given. Since this version is known to be equivalent to the original (4.13), in the remainder of this chapter, $\boldsymbol{\Delta}_{\boldsymbol{\gamma},f}^{(n)}$ will be meant to be (4.14), which is computable from the sample. Finally letting

$$\mathbf{W}_t(\boldsymbol{b}, f) := \frac{[1, b_0 \sum_{i=1}^{p} b_1^{i-1} \prod_{j=1}^{i} F^{-1}(V_{\boldsymbol{b},t-j}^{(n)})]'}{b_0 + b_1 b_0 \sum_{i=1}^{p} b_1^{i-1} \prod_{j=1}^{i} F^{-1}(V_{\boldsymbol{b},t-j}^{(n)})},$$

we have that

$$\overline{\mathbf{W}}^{(n)}(\boldsymbol{\gamma}, f) := \frac{1}{n-p} \sum_{t=p+1}^{n} \mathbf{W}_t(\boldsymbol{\gamma}, f) \xrightarrow{P} \boldsymbol{\mu}_{\mathbf{W}(\boldsymbol{\gamma},f)}, \quad \text{and}$$

$$\frac{1}{n-p} \sum_{t=p+1}^{n} \{\mathbf{W}_t(\boldsymbol{\gamma}, f) - \overline{\mathbf{W}}^{(n)}(\boldsymbol{\gamma}, f)\}\{\mathbf{W}_t(\boldsymbol{\gamma}, f) - \overline{\mathbf{W}}^{(n)}(\boldsymbol{\gamma}, f)\}' \xrightarrow{P} \boldsymbol{\Sigma}_{\mathbf{W}(\boldsymbol{\gamma},f)},$$

for some $\boldsymbol{\mu}_{\mathbf{W}(\boldsymbol{\gamma},f)}$ and $\boldsymbol{\Sigma}_{\mathbf{W}(\boldsymbol{\gamma},f)}$. We are now ready to state the asymptotic representation of the serial linear sign-and-rank statistics $\boldsymbol{\Delta}_{\boldsymbol{\gamma},f}^{(n)}$ of the form (4.14) as follows.

Theorem 4.1 *Let Assumptions 4.1 and 4.2 hold. Then, for any $f \in \mathcal{F}^\tau$ and some suitably large p, we have, under $\boldsymbol{\gamma}$ and g_1,*

$$\begin{aligned}\boldsymbol{\Delta}^{(n)}_{\boldsymbol{\gamma},f} = \frac{1}{\sqrt{n-p}} \sum_{t=p+1}^{n} \Bigg[& \boldsymbol{\psi}_{f,p}[G_1(\xi_{\boldsymbol{\gamma},t}), \ldots, G_1(\xi_{\boldsymbol{\gamma},t-p})] \\ & - E[\boldsymbol{\psi}_{f,p}(U_0, \ldots, U_p) \,|U_0 = G_1(\xi_{\boldsymbol{\gamma},t})] \\ & - \frac{f(1)}{\tau(1-\tau)}(1\{\xi_{\boldsymbol{\gamma},t} \le 1\} - \tau)\boldsymbol{\mu}_{\mathbf{W}(\boldsymbol{\gamma},f)}\Bigg] + o_P(1).\end{aligned} \tag{4.15}$$

If moreover $\mathbf{0} < \int_{[0,1]^{p+1}} |\boldsymbol{\psi}_{f,p}(u_0, u_1, \ldots, u_p)|^{2+\delta} d\boldsymbol{u} < \infty$ *for some* $\delta > 0$*, then, for any* $f \in \mathcal{F}^\tau$*, we have, under* $\boldsymbol{\gamma}$ *and* g_1*,*

$$\boldsymbol{\Delta}^{(n)}_{\boldsymbol{\gamma},f} \xrightarrow{d} N\Big(\mathbf{0},\ \Sigma_f(\boldsymbol{\gamma})\Big), \tag{4.16}$$

$$\text{where} \quad \Sigma_f(\boldsymbol{\gamma}) := \mathcal{I}_f \Sigma_{\mathbf{W}(\boldsymbol{\gamma},f)} + \frac{f^2(1)}{\tau(1-\tau)} \boldsymbol{\mu}_{\mathbf{W}(\boldsymbol{\gamma},f)} \boldsymbol{\mu}'_{\mathbf{W}(\boldsymbol{\gamma},f)}.$$

Proof See Sect. 4.4.1. □

Remark 4.2 Note that the behavior of $\boldsymbol{\Delta}^{(n)}_{\boldsymbol{\gamma},f}$, which is based on the reference density f, is now studied under the true density g_1. Also, due to the invariance properties, the limiting distribution in (4.16) only depends on the reference density f, and not on the true density g_1. ■

To state the "one-step estimator" of $\boldsymbol{\gamma}$, the asymptotic behavior under $\boldsymbol{\gamma}$ and g of $\boldsymbol{\Delta}^{(n)}_{\tilde{\boldsymbol{\gamma}}_n,f}$, for some local alternative $\tilde{\boldsymbol{\gamma}}_n$ defined below, is needed. For this purpose, let us denote, in regard to (4.15),

$$\boldsymbol{T}^{(n)}_{\boldsymbol{b},f} := \frac{1}{\sqrt{n-p}} \sum_{t=p+1}^{n} \begin{pmatrix} \varphi_f[F^{-1}(G_1(\xi_{\boldsymbol{b},t}))] \left(\boldsymbol{W}_t(\boldsymbol{b}, f) - \boldsymbol{\mu}_{\mathbf{W}(\boldsymbol{b},f)}\right) \\ 1\{\xi_{\boldsymbol{b},t} \le 1\} - \tau \end{pmatrix}.$$

Now, consider a local alternative, with respect to $\boldsymbol{\gamma}$, of the form $\boldsymbol{\gamma}_n = \boldsymbol{\gamma} + \boldsymbol{h}_n/\sqrt{n} + o(n^{-1/2})$, with $\boldsymbol{h}_n \to \boldsymbol{h}$ as $n \to \infty$. Note that this is still within our ULAN framework. Then, under the following assumption, we have the asymptotic linearity.

Assumption 4.3

(i) $\sup_{x \in \mathbb{R}_+} |x g_1(x)| < \infty$ and $\lim_{x \to \infty} x g_1(x) = 0$;

(ii) $\varphi^*_f := \varphi_f \circ F^{-1}$ is nondecreasing and differentiable with its derivative $\dot{\varphi}^*_f$ being uniformly continuous on [0, 1].

Lemma 4.1 *Let Assumptions 4.1 (with Remark 4.1) and 3 hold. Then, for an arbitrary sequence* $\boldsymbol{\gamma}_n = \boldsymbol{\gamma} + \boldsymbol{h}_n/\sqrt{n} + o(n^{-1/2})$ *with* $\boldsymbol{h}_n \to \boldsymbol{h}$, *we have*

$$\boldsymbol{T}^{(n)}_{\boldsymbol{\gamma}_n, f} - \boldsymbol{T}^{(n)}_{\boldsymbol{\gamma}, f} + \begin{pmatrix} \mathcal{I}_{fg_1} \Sigma_{1,\boldsymbol{W}(\boldsymbol{\gamma},f)} \\ -g_1(1)\boldsymbol{\mu}'_{\boldsymbol{W}(\boldsymbol{\gamma})} \end{pmatrix} \boldsymbol{h} = o_P(1), \tag{4.17}$$

where $\mathcal{I}_{fg_1} := \int_0^1 \varphi_f[F^{-1}(u)]\varphi_{g_1}[G_1^{-1}(u)]du$, $\boldsymbol{\mu}_{\boldsymbol{W}(\boldsymbol{\gamma})} := E[\boldsymbol{W}_t(\boldsymbol{\gamma})]$, *and* $\Sigma_{1,\boldsymbol{W}(\boldsymbol{\gamma},f)}$ *is such that*

$$S^{(n)}_{1,\boldsymbol{W}(\boldsymbol{\gamma},f)} := \frac{1}{n-p} \sum_{t=p+1}^{n} \boldsymbol{W}_t(\boldsymbol{\gamma}, f)\boldsymbol{W}_t(\boldsymbol{\gamma})' \xrightarrow{P} \Sigma_{1,\boldsymbol{W}(\boldsymbol{\gamma},f)}. \tag{4.18}$$

Proof The proof for the first component of (4.17) is similar to that of Koul (1992, Theorem 7.3b.1). So we omit the details (See Taniai (2009)) and only remark that the quantity Q therein corresponds here to

$$\begin{aligned} Q &\equiv \int_0^1 G_1^{-1}(u) g_1(G_1^{-1}(u))\dot{\varphi}^*_f(u)du \\ &= \left[G_1^{-1}(u) g_1(G_1^{-1}(u))\varphi^*_f(u) \right]_0^1 - \int_0^1 \frac{d\{G_1^{-1}(u)g_1(G_1^{-1}(u))\}}{du} \varphi^*_f(u)du \\ &= \int_0^1 \left\{ -1 - \frac{G_1^{-1}(u)g_1'(G_1^{-1}(u))}{g_1(G_1^{-1}(u))} \right\} \varphi^*_f(u)du = \mathcal{I}_{fg_1}, \end{aligned}$$

by Assumption 4.3-(i). Also, the second component of (4.17) follows from an application of Koul (1992, Corollary 7.2.2). □

Finally in preparing for the one-step estimator, a "discretization trick" is to be introduced.

Definition 4.1 For any sequence of estimators $\hat{\theta}_n$, the discretized estimator $\bar{\theta}$ is defined to be the nearest vertex of $\{\theta : \theta = \frac{1}{\sqrt{n}}(i_1, i_2, ..., i_k)', i_j : \text{integers}\}$.

We denote by $\bar{\boldsymbol{\gamma}}_n\ (= \bar{\boldsymbol{\gamma}}_n(\tau))$ a locally discretized version of $\hat{\boldsymbol{\gamma}}_n(\tau)$ which is defined at (4.3). Recall here that Corollary 4.1 ensures the $\sqrt{n}$-consistency of $\hat{\boldsymbol{\gamma}}_n(\tau)$. So, if we write as $\hat{\boldsymbol{\gamma}}_n(\tau) = \boldsymbol{\gamma} + \boldsymbol{k}_n/\sqrt{n}$ then $\boldsymbol{k}_n$ is a sequence which is bounded in probability. The great advantage of the discritization is that it allows us to regard $\boldsymbol{k}_n$ further as a nonrandom sequence (See Kreiss (1987, Lemma 4.4), or Linton (1993) etc.), i.e., $\bar{\boldsymbol{\gamma}}_n = \boldsymbol{\gamma} + \boldsymbol{h}_n/\sqrt{n}$ as in Lemma 4.1.

Now, using the rank-based central sequence $\underset{\sim}{\boldsymbol{\Delta}}^{(n)}_{\bar{\boldsymbol{\gamma}}_n, f}$ based on a $\sqrt{n}$-consistent discretized estimator $\bar{\boldsymbol{\gamma}}_n$, we construct some $\tilde{\boldsymbol{\gamma}}_n$ for which $\sqrt{n}(\tilde{\boldsymbol{\gamma}}_n - \boldsymbol{\gamma})$ behaves well. If $\sqrt{n}(\tilde{\boldsymbol{\gamma}}_n - \bar{\boldsymbol{\gamma}}_n)$ is of the form $A_n^{-1}\underset{\sim}{\boldsymbol{\Delta}}^{(n)}_{\bar{\boldsymbol{\gamma}}_n, f}$ for some matrix A_n, then we have by (4.17), under $\boldsymbol{\gamma}$ and g_1,

$$\sqrt{n}(\tilde{\boldsymbol{\gamma}}_n - \boldsymbol{\gamma}) = \sqrt{n}(\tilde{\boldsymbol{\gamma}}_n - \bar{\boldsymbol{\gamma}}_n) + \sqrt{n}(\bar{\boldsymbol{\gamma}}_n - \boldsymbol{\gamma}) = A_n^{-1}\boldsymbol{\Delta}^{(n)}_{\bar{\boldsymbol{\gamma}}_n, f} + \boldsymbol{h}_n$$
$$= A_n^{-1}\left(\boldsymbol{\Delta}^{(n)}_{\boldsymbol{\gamma}, f} - \mathcal{I}_{fg_1}\Sigma_{1,\mathbf{W}(\boldsymbol{\gamma},f)}\boldsymbol{h} - \frac{f(1)g_1(1)}{\tau(1-\tau)}\boldsymbol{\mu}_{\mathbf{W}(\boldsymbol{\gamma},f)}\boldsymbol{\mu}'_{\mathbf{W}(\boldsymbol{\gamma})}\boldsymbol{h}\right) + \boldsymbol{h}_n + o_P(1). \tag{4.19}$$

This, together with $\boldsymbol{h}_n \to \boldsymbol{h}$, suggests us to choose A_n to be consistent for

$$\Sigma_{fg_1}(\boldsymbol{\gamma}) := \mathcal{I}_{fg_1}\Sigma_{1,\mathbf{W}(\boldsymbol{\gamma},f)} + \frac{f(1)g_1(1)}{\tau(1-\tau)}\boldsymbol{\mu}_{\mathbf{W}(\boldsymbol{\gamma},f)}\boldsymbol{\mu}'_{\mathbf{W}(\boldsymbol{\gamma})}, \tag{4.20}$$

so that (4.19) becomes asymptotically equivalent to $A_n^{-1}\boldsymbol{\Delta}^{(n)}_{\boldsymbol{\gamma},f}$. For this purpose, let us remind (4.18) and define

$$M^{(n)}_{1,\mathbf{W}(\boldsymbol{\gamma},f)} := \left(\frac{1}{n-p}\sum_{t=p+1}^{n}\mathbf{W}_t(\boldsymbol{\gamma},f)\right)\left(\frac{1}{n-p}\sum_{t=p+1}^{n}\mathbf{W}_t(\boldsymbol{\gamma})\right)'.$$

The consistencies $S^{(n)}_{1,\mathbf{W}(\bar{\boldsymbol{\gamma}}_n,f)} \xrightarrow{P} \Sigma_{1,\mathbf{W}(\boldsymbol{\gamma},f)}$ and $M^{(n)}_{1,\mathbf{W}(\bar{\boldsymbol{\gamma}}_n,f)} \xrightarrow{P} \boldsymbol{\mu}_{\mathbf{W}(\boldsymbol{\gamma},f)}\boldsymbol{\mu}'_{\mathbf{W}(\boldsymbol{\gamma})}$ can be verified in the manner of Linton (1993, Theorem 2). Now, we define the one-step estimator $\tilde{\boldsymbol{\gamma}}_n$ as follows.

Definition 4.2 The rank-based one-step estimator of $\boldsymbol{\gamma}$ starting from the $\sqrt{n}$-consistent and locally discrete estimator $\bar{\boldsymbol{\gamma}}_n$ and based on reference density $f \in \mathcal{F}^\tau$ is defined as

$$\tilde{\boldsymbol{\gamma}}_{n,f} := \bar{\boldsymbol{\gamma}}_n + \hat{\boldsymbol{\Sigma}}_{fg_1}(\bar{\boldsymbol{\gamma}}_n)^{-1}\frac{\boldsymbol{\Delta}^{(n)}_{\bar{\boldsymbol{\gamma}}_n,f}}{\sqrt{n}}, \qquad \text{with} \tag{4.21}$$
$$\hat{\boldsymbol{\Sigma}}_{fg_1}(\bar{\boldsymbol{\gamma}}_n) := \hat{\mathcal{I}}_{fg_1}S^{(n)}_{1,\mathbf{W}(\boldsymbol{\gamma},f)} + \frac{\tau}{1-\tau}\cdot\frac{f(1)}{-\tau}\cdot\hat{\mu}_{\varphi_{g_1},L}M^{(n)}_{1,\mathbf{W}(\boldsymbol{\gamma},f)},$$

where $\hat{\mathcal{I}}_{fg_1}$ and $\hat{\mu}_{\varphi_{g_1},L}$ are consistent estimates of $\mathcal{I}_{fg_1}$ and

$$\mu_{\varphi_{g_1},L} := E[\varphi_{g_1}[G_1^{-1}(U)]\,|U \leq \tau] \quad (= -g_1(1)/\tau),$$

respectively.

The consistent estimators $\hat{\mathcal{I}}_{fg_1}$ and $\hat{\mu}_{\varphi_{g_1},L}$ can be obtained in the manner of Hallin et al. (2006a, Sect. 4.2), which will be done without the kernel estimation of g_1. (See Taniai (2009) for the details.)

If Assumptions 4.1 and 4.2 hold then we have, under $\boldsymbol{\gamma}$ and g_1,

$$\sqrt{n}(\tilde{\boldsymbol{\gamma}}_{n,f} - \boldsymbol{\gamma}) \xrightarrow{d} N\left(\mathbf{0},\ \Sigma^{-1}_{fg_1}(\boldsymbol{\gamma})\Sigma_f(\boldsymbol{\gamma})\Sigma^{-1}_{fg_1}(\boldsymbol{\gamma})\right), \tag{4.22}$$

as a consequence of Theorem 4.1. Also, similarly to Hallin et al. (2008), this $\tilde{\boldsymbol{\gamma}}_{n,f}$ is shown to attain the semiparametric lower bound.

Proposition 4.2 *Under Assumptions 4.1–4.3, the one-step estimator $\tilde{\boldsymbol{\gamma}}_{n,f}$ defined by (4.21) for $\boldsymbol{\gamma}$ is semi-parametrically efficient at $f = g_1$.*

Proof The proof is similar to that of Hallin et al. (2006b, Lemma 2), i.e., the parametric submodels are constructed as in Bickel et al. (1998, Example 3.2.1). We omit the details. □

4.2.3 Numerical Studies

Based on the result of Lemma 4.1 and Le Cam's third lemma (See van der Vaart (1998)), we can have the convergence-in-law of $\boldsymbol{T}^{(n)}_{\boldsymbol{\gamma}_0}$ under the local alternatives $\boldsymbol{\gamma}_n = \boldsymbol{\gamma}_0 + \boldsymbol{h}/\sqrt{n}$ (and the "true" density g_1). Namely, we have, under the local alternatives $\boldsymbol{\gamma}_n$,

$$\boldsymbol{\Delta}^{(n)}_{\boldsymbol{\gamma}_0, f} \xrightarrow{d} N\left(\Sigma_{fg_1}(\boldsymbol{\gamma}_0)\boldsymbol{h},\ \Sigma_f(\boldsymbol{\gamma}_0)\right), \qquad \text{under} P_{\boldsymbol{\gamma}_n, g_1} \tag{4.23}$$

and hence the distribution (under $\boldsymbol{\gamma}_n$) of Lagrange multiplier (LM) statistics

$$LM^{(n)}_f := \boldsymbol{\Delta}^{(n)\prime}_{\boldsymbol{\gamma}_0, f} \Sigma^{-1}_f(\boldsymbol{\gamma}_0)\boldsymbol{\Delta}^{(n)}_{\boldsymbol{\gamma}_0, f} \tag{4.24}$$

is asymptotically a noncentral chi-square with 2 degrees of freedom and noncentrality parameter

$$\boldsymbol{h}'\Sigma'_{fg_1}(\boldsymbol{\gamma}_0)\Sigma^{-1}_f(\boldsymbol{\gamma}_0)\Sigma_{fg_1}(\boldsymbol{\gamma}_0)\boldsymbol{h}. \tag{4.25}$$

Note that the distribution of $LM^{(n)}_f$ under the null $\boldsymbol{\gamma}_0$ is asymptotically a (central) chi-square with 2 degrees of freedom, χ^2_2. Thus, the tests that reject the null hypothesis $H_0 : \boldsymbol{\gamma} = \boldsymbol{\gamma}_0$ if $LM^{(n)}_f$ exceeds $F^{-1}_{\chi^2_2}(1-\alpha)$ are asymptotically of level α.

Assume now that we have two such level α tests $A^{(1)}$ and $A^{(2)}$ based on the statistics $LM^{(n_1)}_{f_1}$ and $LM^{(n_2)}_{f_2}$. Also, let us assume that their noncentrality parameter (under g_1) can be expressed in a quadratic form of $\boldsymbol{h}$, and denote them as $\lambda_{g_1}(f_i, \boldsymbol{h}) = \boldsymbol{h}'B_i\boldsymbol{h}$, $i = 1, 2$. Then the two tests have the same limiting power against the same sequence of alternatives if

$$\frac{\boldsymbol{h}_1}{\sqrt{n_1}} = \frac{\boldsymbol{h}_2}{\sqrt{n_2}} \tag{4.26}$$

and

$$\lambda_{g_1}(f_1, \boldsymbol{h}_1) = \lambda_{g_1}(f_2, \boldsymbol{h}_2). \tag{4.27}$$

So, from (4.26) and (4.27), we see that the asymptotic relative efficiency (ARE, Pitman efficiency) of $A^{(2)}$ with respect to $A^{(1)}$ under g_1 should be

$$\mathrm{ARE}_{g_1}(A^{(2)}, A^{(1)}) := \lim \frac{n_1}{n_2} = \frac{\lambda_{g_1}(f_2, \boldsymbol{h}_1)}{\lambda_{g_1}(f_1, \boldsymbol{h}_1)}. \tag{4.28}$$

Since this involves $\boldsymbol{h}_1 = (h_{11}, h_{12})'$, the unique answer regarding ARE of $A^{(2)}$ w.r.t. $A^{(1)}$ may not be possible. However, as suggested in Puri and Sen (1971, Sect. 3.8.3), the supremum and the infinimum of $\mathrm{ARE}_{g_1}(A^{(2)}, A^{(1)})$ can be obtained as the largest and smallest eigenvalues of $B_2 B_1^{-1}$ by the Courant-Fischer minimax theorem. We will show these supremum and infinimum later. Still, it is worthwhile, in fact, to observe also the values of (4.28) for $\boldsymbol{h}_1 = (1, 0)'$ and for $\boldsymbol{h}_1 = (0, 1)'$. This is because they would correspond to the testing, not both but, one of the parameters while keeping the other parameter fixed. In order to make those displays, further description of our setting is now explained as follows.

Let us first define the reference and/or underlying densities we consider. Below, we see some numerical results concerning ARE of LM tests for a family of models, which contains the squared series of ARCH model (4.1). As ARCH models require the (unsquared) innovation Z_t to be i.i.d.(0, 1), let us think of generating the density of squared innovations from such i.i.d.(0, 1) densities. So, let us suppose that $\{Z_t\}$ is a sequence of i.i.d.(0, 1) random variables with continuous distribution function F_Z of the symmetric density f_Z. Then

$$F_{Z^2}(z) = P\{Z_t^2 \le z\} = \begin{cases} 0, & z \le 0, \\ 2F_Z(\sqrt{z}) - 1, & z > 0. \end{cases} \tag{4.29}$$

We shall now compute (4.29) in the following particular choices of the innovation distribution function F_Z, and define families of densities, which contain such particular choices within.

(i) F_Z being Normal:

$$F_{\mathrm{N}}(z) = \Phi(z) = \int_{-\infty}^{z} \phi(t)dt = \frac{1}{\sqrt{2\pi}} \int_{-\infty}^{z} e^{-t^2/2} dt,$$
$$\Rightarrow \quad F_{\mathrm{N}^2}(z) = 2\Phi(\sqrt{z}) - 1, \qquad f_{\mathrm{N}^2}(z) = \frac{\phi(\sqrt{z})}{\sqrt{z}}.$$

In this case, we have $f_{\mathrm{N}^2} \in \mathcal{F}^{0.6827}$ so that the commonly performed analysis of $\boldsymbol{\beta}$ corresponds to the situation $\tau = 0.6827$. But our analysis of $\boldsymbol{\gamma}$ $(= \boldsymbol{\gamma}(\tau))$ concerns more generally the class of the following densities for $\tau \in (0, 1)$:

$$f_{\mathrm{N}^2}^{\tau}(z) := f_{\mathrm{N}^2}(z \cdot F_{\mathrm{N}^2}^{-1}(\tau)) \cdot F_{\mathrm{N}^2}^{-1}(\tau) \left(= \frac{dF_{\mathrm{N}^2}(z \cdot F_{\mathrm{N}^2}^{-1}(\tau))}{dz} \right) \in \mathcal{F}^{\tau}.$$

Table 4.1 The values of $(E[\xi_0^2])^{-1/2}$: In order to maintain the stationarity of the process, γ_1 $(= \gamma_1(\tau))$ should be smaller than these values

		τ								
		0.15	0.25	0.35	0.45	0.55	0.65	0.75	0.85	0.95
g_1	$f_{N^2}^{\tau}$	0.02	0.06	0.12	0.20	0.33	0.50	0.76	1.20	2.22
	$f_{DE^2}^{\tau}$	0.005	0.02	0.04	0.07	0.13	0.22	0.39	0.73	1.83
	$f_{LGT^2}^{\tau}$	0.01	0.04	0.08	0.14	0.23	0.36	0.56	0.94	1.99

(ii) F_Z being Double Exponential (bilateral exponential, Laplace):

$$F_{\text{DE}}(z) = \text{DE}\left(z; 0, \frac{1}{\sqrt{2}}\right) = \frac{\sqrt{2}}{2} \int_{-\infty}^{z} e^{-\sqrt{2}|t|} dt,$$
$$\Rightarrow \quad f_{\text{DE}^2}(z) = \frac{1}{\sqrt{2z}} \exp(-\sqrt{2z}).$$

In this case, while $f_{\text{DE}^2} \in \mathcal{F}^{0.7569}$, we apply the class of densities

$$f_{\text{DE}^2}^{\tau}(z) := f_{\text{DE}^2}(z \cdot F_{\text{DE}^2}^{-1}(\tau)) \cdot F_{\text{DE}^2}^{-1}(\tau) \in \mathcal{F}^{\tau},$$

(iii) F_Z being Logistic:

$$F_{\text{LGT}}(z) = \text{Logistic}\left(z; 0, \frac{\sqrt{3}}{\pi}\right) = \left\{1 + \exp\left(-\frac{\pi}{\sqrt{3}} z\right)\right\}^{-1},$$
$$\Rightarrow \quad f_{\text{LGT}^2}(z) = \frac{\pi}{\sqrt{3z}} \frac{\exp(-\sqrt{\pi^2 z/3})}{\{1 + \exp(-\sqrt{\pi^2 z/3})\}^2}.$$

Again, while $f_{\text{LGT}^2} \in \mathcal{F}^{0.7196}$, we apply

$$f_{\text{LGT}^2}^{\tau}(z) := f_{\text{LGT}^2}(z \cdot F_{\text{LGT}^2}^{-1}(\tau)) \cdot F_{\text{LGT}^2}^{-1}(\tau) \in \mathcal{F}^{\tau}.$$

Being provided the exact form of $g_1 \in \mathcal{F}^{\tau}$, we have the collection of γ_1, which makes the process stationary. According to the sufficient condition of Remark 4.1, the following Table 4.1 tells that the availability of γ_1 are limited especially for small τ.

Now, we start with calculating ARE of rank based LM statistics (4.24) with respect to a LM statistic which is based on the parametric "Gaussian" (: more accurately, squared Gaussian $f_{N^2}^{\tau}$) score function. In order to ensure the asymptotic results for the latter quantity, the Gaussian parametric score, here we suppose that the underlying density is being correctly specified, i.e., $g_1 = f_{N^2}^{\tau}$. Consequently, our rank-based tests, or any tests, can not outperform such a correctly specified parametric score test. But still we use this as a benchmark, and observe the performances of several

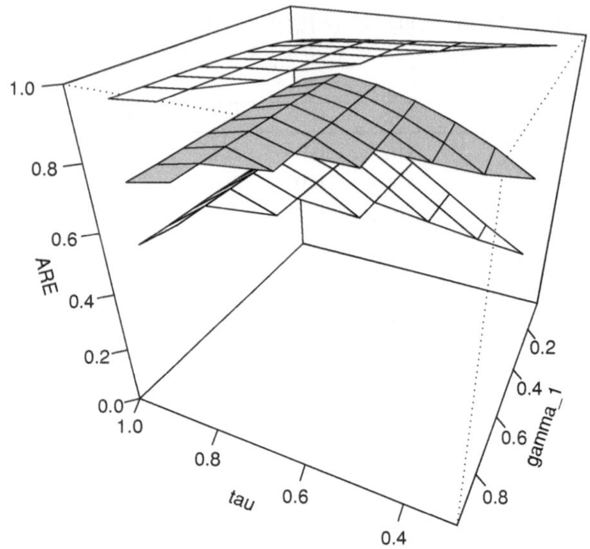

Fig. 4.1 The *top* and the *bottom* surfaces are sup and inf of $e_{g_1}(\mathrm{RB}(f^{\tau}_{\mathrm{N}^2}), \mathrm{P}(g_1))$, the ARE of rank-based test (with reference density $f^{\tau}_{\mathrm{N}^2}$) w.r.t. the correctly-specified parametric score test under the true density g_1. Those sup and inf are taken among all the direction $\boldsymbol{h}_1$ of the local perturbation. (The values for a specific direction $\boldsymbol{h}_1 = (0, 1)'$ is provided as the shadowed surface.) In this case, our setting $g_1 = f^{\tau}_{\mathrm{N}^2}$ means the equality of the reference density and the true density. Still, we may observe that the supremum of ARE seems to be not attaining 1

reference densities against this benchmark. To summarize, the quantity which we first concern is

$$\begin{aligned}
&\mathrm{ARE}_{g_1}(\text{Rank-Based with} f, \text{Parametric with } g_1)\\
&= \frac{\boldsymbol{h}'\Sigma'_{fg_1}(\boldsymbol{\gamma}_0)\Sigma_f^{-1}(\boldsymbol{\gamma}_0)\Sigma_{fg_1}(\boldsymbol{\gamma}_0)\boldsymbol{h}}{\boldsymbol{h}'(\mathcal{I}_{g_1}(\Sigma_{\mathbf{W}(\boldsymbol{\gamma},g_1)} + M_{\mathbf{W}(\boldsymbol{\gamma},g_1)}))\boldsymbol{h}}\\
&\overset{(\text{say})}{=} e_{g_1}(\mathrm{RB}(f), \mathrm{P}(g_1)),
\end{aligned} \tag{4.30}$$

with $g_1 = f^{\tau}_{\mathrm{N}^2}$ and several f are chosen for comparison. As discussed in (4.28), we can calculate the supremum and infinimum of $e_{g_1}(\mathrm{RB}(f), \mathrm{P}(g_1))$. Figure 4.1 shows the case of $f \equiv f^{\tau}_{\mathrm{N}^2}$, and there the values of $e_{g_1}(\mathrm{RB}(f), \mathrm{P}(g_1))$ for $\boldsymbol{h}_1 = (0, 1)'$ are also shown. There, the sample size is $n = 500$ and the finite approximation is done with $p = 50$. Also, there we set γ_0 to be $1 - \gamma_1$ in order to keep the unconditional variance being the same regardless of the changes of ARCH parameters (this does not make any substantial effect on the result).

In fact, the cases where $f \equiv f^{\tau}_{\mathrm{DE}^2}$ and $f \equiv f^{\tau}_{\mathrm{LGT}^2}$ yield the similar result as shown in Fig. 4.1 (See Taniai (2009)). According to this observation, it seems that the value of the parameter γ_1 does not have substantial effect on the result. On the other hand, the dependence on τ is evident. Those τ's being too small or too big would makes our rank-based inference difficult comparing to the parametric method. Also, even for those supremum values, they seem to be not attaining 1. This means that there is a definite gap between the semiparametric efficiency (the best feasible solution), and the correctly specified parametric efficiency (the best possible but infeasible solution).

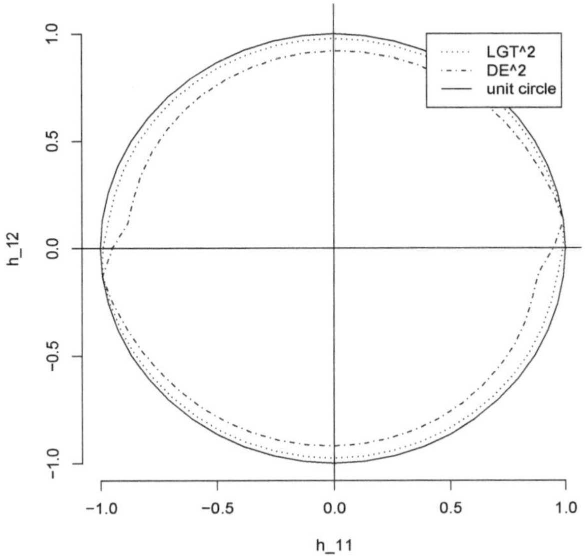

Fig. 4.2 The ARE among rank based tests, $e_{g_1}(\mathrm{RB}(f), \mathrm{RB}(g_1))$ with $g_1 = f^{\tau}_{\mathrm{N}^2}$ for $\tau = 0.35$ and $\gamma_1 = 0.1$. Since here the underlying density is $g_1 = f^{\tau}_{\mathrm{N}^2}$, the unit circle represents the rank based test with the reference density based on Gaussian

Next, let us have a close look to the ARE among those rank based procedures. That is, we observe the value of

$$\begin{aligned}
&\mathrm{ARE}_{g_1}(\text{Rank-Based with } f, \text{Rank-Based with } g_1) \\
&= \frac{\boldsymbol{h}'\Sigma'_{fg_1}(\boldsymbol{\gamma}_0)\Sigma_f^{-1}(\boldsymbol{\gamma}_0)\Sigma_{fg_1}(\boldsymbol{\gamma}_0)\boldsymbol{h}}{\boldsymbol{h}'\Sigma_{g_1}(\boldsymbol{\gamma}_0)\boldsymbol{h}} \\
&\overset{\text{(say)}}{=} e_{g_1}(\mathrm{RB}(f), \mathrm{RB}(g_1)),
\end{aligned} \tag{4.31}$$

with several choices of f and g_1 for comparison. Figures 4.2 and 4.3 show $e_{g_1}(\mathrm{RB}(f), \mathrm{RB}(g_1))$ for each direction $\boldsymbol{h}_1 = (h_{11}, h_{12})'$ with $g_1 = f^{\tau}_{\mathrm{N}^2}$ for different choices of τ and γ_1. This explain how much of efficiency will be lost according to our choices for the reference densities, under the case where true density $g_1 = f^{\tau}_{\mathrm{N}^2}$ (i.e., Gaussian-based case). As for the other cases, i.e., Double Exponential based case ($g_1 \equiv f^{\tau}_{\mathrm{DE}^2}$) and Logistic based case ($g_1 \equiv f^{\tau}_{\mathrm{LGT}^2}$), we refer to Taniai (2009).

According to the figures, it can be seen that the rank-based tests are robust against the misspecification of the underlying density. So, we may conclude as follows: Once we enter the semi-parametrics, the rank-based inference in this chapter, the price for misspecification can be reduced. Also, the reason why we may need to enter is due to the fact that we accepted, in the first place, our ignorance or indifference about the underlying densities.

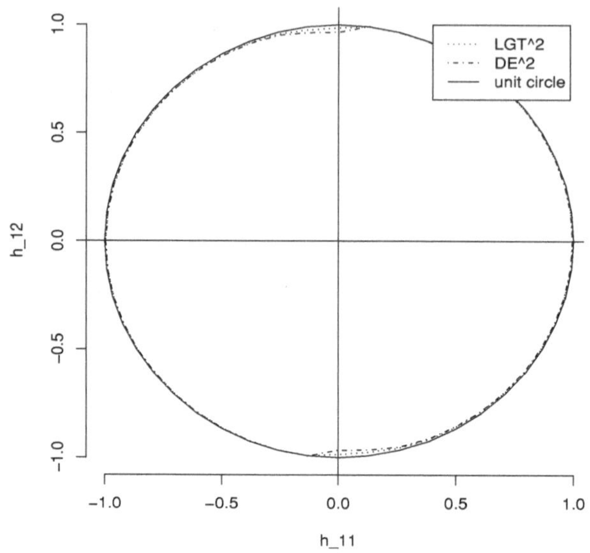

Fig. 4.3 (The same as Figure 4.2, but for $\tau = 0.85$ and $\gamma_1 = 0.9$.)

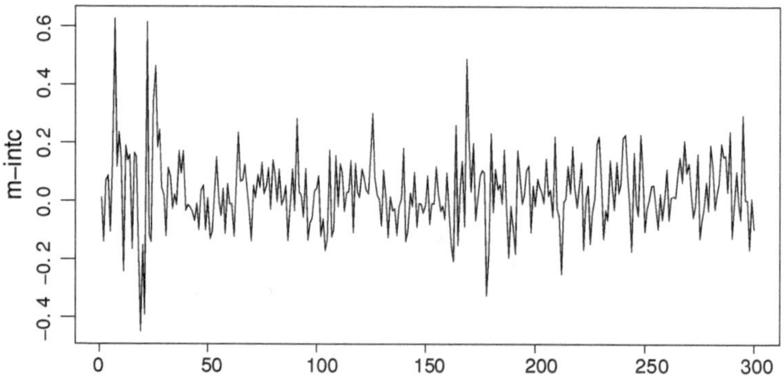

Fig. 4.4 Time plot of the monthly log stock returns of Intel from January 1973 to December 1997

Finally, let us see the real data application with the monthly log stock returns of Intel Corporation from January 1973 to December 1997, which consists of $n = 300$ data points (Fig. 4.4). This data can be found in Tsay (2002, Example 3.1), and studied to behave as ARCH(1) under the Gaussian assumption. For example, denoting X_t for the squared value of log stock returns at time t, ordinary Quantile Regression analysis for $\tau = 0.7$ becomes,

$$\hat{F}_{X_t}^{-1}(0.7|x_{t-1}) = 0.0137838 + 0.1379812x_{t-1}.$$

Table 4.2 MSE of conditional quantiles: This table shows, for each cases of g_1, how the ordinary QR estimator (the row "QR") and our rank-based estimators (the row "$\mathrm{OS}_{\mathrm{N}^2}$", "$\mathrm{OS}_{\mathrm{DE}^2}$", "$\mathrm{OS}_{\mathrm{LGT}^2}$") performs in estimation

		τ				
		0.1	0.3	0.5	0.7	0.9
$g_1 = f^{\tau}_{\mathrm{N}^2}$	QR	5.00×10^{-6}	1.49×10^{-4}	2.11×10^{-3}	7.41×10^{-3}	5.73×10^{-2}
	$\mathrm{OS}_{\mathrm{N}^2}$	14.9×10^{-6}	1.44×10^{-4}	1.12×10^{-3}	4.67×10^{-3}	4.89×10^{-2}
	$\mathrm{OS}_{\mathrm{DE}^2}$	14.9×10^{-6}	1.45×10^{-4}	1.13×10^{-3}	4.77×10^{-3}	5.12×10^{-2}
	$\mathrm{OS}_{\mathrm{LGT}^2}$	14.9×10^{-6}	1.46×10^{-4}	1.14×10^{-3}	4.77×10^{-3}	5.01×10^{-2}
$g_1 = f^{\tau}_{\mathrm{DE}^2}$	QR	9.91×10^{-6}	2.95×10^{-4}	4.19×10^{-3}	14.6×10^{-3}	11.3×10^{-2}
	$\mathrm{OS}_{\mathrm{N2}}$	15.2×10^{-6}	1.18×10^{-4}	1.10×10^{-3}	5.46×10^{-3}	11.0×10^{-2}
	$\mathrm{OS}_{\mathrm{DE2}}$	15.2×10^{-6}	1.19×10^{-4}	1.10×10^{-3}	5.38×10^{-3}	10.5×10^{-2}
	$\mathrm{OS}_{\mathrm{LGT2}}$	15.1×10^{-6}	1.18×10^{-4}	1.09×10^{-3}	5.39×10^{-3}	10.6×10^{-2}

QR corresponds to ordinary quantile regression estimator (4.3), and OS_f corresponds to our rank-based one-step estimator (4.21) with reference density f and $p = 5$. *N* Normal, *DE* Double Exponential, *LGT* Logistic. (The case $g_1 = f^{\tau}_{\mathrm{LGT}^2}$ is similar to that of $f^{\tau}_{\mathrm{DE}^2}$ so we omitted.) For example, the category "$g_1 = f^{\tau}_{\mathrm{N}^2}$" exhibits the case where the simulation is generated from the density $f^{\tau}_{\mathrm{N}^2}$, and hence the estimators "$\mathrm{OS}_{\mathrm{DE2}}$" and "$\mathrm{OS}_{\mathrm{LGT2}}$" must be misspecifying in fact. As a consequence their performances are worse than "$\mathrm{OS}_{\mathrm{N2}}$," but we note that even these misspecifying estimators can perform better than the ordinary QR estimator

In order to check the efficiency gain of our one-step estimator, we compute the mean squared errors (MSE) of the conditional quantile of the squared series. That is, we compute

$$\begin{aligned}
\mathrm{MSE}^{(\mathrm{QR})} &:= E\sqrt{n}\{\hat{F}^{-1}_{X_t}(\tau|x_{t-1}) - F^{-1}_{X_t}(\tau|x_{t-1})\}^2 \\
&= E[\{\sqrt{n}(\boldsymbol{X}'_{t-1}(\hat{\boldsymbol{\gamma}}_n(\tau) - \boldsymbol{\gamma}(\tau))\}^2] \\
&\approx \frac{1}{n}\sum_{t=1}^{n} \mathrm{tr}[\boldsymbol{X}_{t-1}\boldsymbol{X}'_{t-1}\frac{\tau(1-\tau)}{g_1^2(1)} D^{-1}_{1,\hat{\boldsymbol{\gamma}}_n(\tau)} D_0 D^{-1}_{1,\hat{\boldsymbol{\gamma}}_n(\tau)}]
\end{aligned}$$

and

$$\begin{aligned}
\mathrm{MSE}^{(\mathrm{OS})}_f &= E[\{\sqrt{n}(\boldsymbol{X}'_{t-1}(\tilde{\boldsymbol{\gamma}}_{n,f}(\tau) - \boldsymbol{\gamma}(\tau))\}^2] \\
&\approx \frac{1}{n}\sum_{t=1}^{n} \mathrm{tr}[\boldsymbol{X}_{t-1}\boldsymbol{X}'_{t-1}\Sigma^{-1}_{fg_1}(\tilde{\boldsymbol{\gamma}}_{n,f}(\tau))\Sigma_f(\tilde{\boldsymbol{\gamma}}_{n,f}(\tau))\Sigma^{-1}_{fg_1}(\tilde{\boldsymbol{\gamma}}_{n,f}(\tau))].
\end{aligned}$$

The computed MSEs for several choice of reference density f and the true density g_1 are shown in Table 4.2. Note that, there is no way to find the true density g_1 since this is the real data.

From Table 4.2, we find that our rank-based one-step estimator improves efficiency for moderate choice of τ. Unfortunately, for very small τ (e.g., $\tau = 0.1$) our estimator failed to improve efficiency. This may be related with the remark from Table 4.1 since the estimated value is found to be $\hat{\gamma}_{n,1}(0.1) = 0.012$. That is, our efficiency gain is realized only when the model (4.5) maintains the stationarity, not only the model

(4.2). Still, it is remarkable that, for moderate value of τ, the improvement method of ours is achieved even without the knowledge of true density g_1. In addition, the robustness against the misspecification (as noted in Figs. 4.2 and 4.3) is observed in this analysis as well.

4.3 Asymptotics of Realized Volatility with Non-Gaussian ARCH(∞) Microstructure Noise

In order to control the risk of financial assets, good estimation of the conditional variance, volatility, is indispensable. For this purpose, the quantity known as RV is having attention, especially when the frequency of the sampling is high (cf. Dacorogna et al. (2001)). Namely, let X_t denote the log-price of an asset, and follow some process such as $dX_t = \mu_t dt + \sigma_t dW_t$ which we will discuss later in (4.35) of Sect. 4.3.1. Then the RV (or sometimes called also as Realized Variance) over time $[0, T]$ is defined as the sum of the frequently sampled squared returns (see Andersen et al. (2003); McAleer and Medeiros (2008)):

$$[X, X]_T := \sum_{t=1}^{T} (X_t - X_{t-1})^2. \tag{4.32}$$

This quantity estimates the Integrated Volatility (IV) defined as

$$\langle X, X \rangle_T := \int_0^T \sigma_t^2 dt, \tag{4.33}$$

which corresponds to the quadratic variation in the Stochastic Volatility model (see Barndorff-Nielsen and Shephard (2002)).

But here we need to pay attention also to the existence of market microstructure. That is, it is known that the estimator $[X, X]_T$ does not converge as sampling frequency increases when there is market microstructure dynamics behind. Here, the market microstructure can arise through the bid-ask bounce, asynchronous trading, infrequent trading, and price discreteness, among other factors (see Madhavan (2000); O'Hara (1997)). Still, Zhang et al. (2005) proposed some subsample-based estimator to estimate the integrated volatility and proved its consistency in the presence of i.i.d. market microstructure noise. Their estimator is defined by adjusting the bias from the following preliminary estimator:

$$[X, X]_T^{(\text{avg})} := \frac{1}{K} \sum_{k=1}^{K} [X, X]_T^{(k)} \tag{4.34}$$

where

$$[X, X]_T^{(k)} := \sum_{t=1}^{T_k} \left(X_{k-1+tK} - X_{k-1+(t-1)K} \right)^2, \qquad \text{with}$$

$$T_k := \lfloor \frac{T-k+1}{K} \rfloor. \qquad (\lfloor x \rfloor \text{ is the integer part of a real number } x)$$

Here, K is prespecified and the asymptotics are studied under suitably chosen T and K, as we will see also in Theorem 4.3 below. Namely, this estimator $[X, X]_T^{(\text{avg})}$ firstly uses the time scale, for example, 10 min ($K = 10$) to measure from 9:30 with the intervals 9:30–9:40, 9:40–9:50, ..., then 9:31–9:41, 9:41–9:51, and so on. Then those will be averaged so that we make use of the full data whose time scale is, say, 1 min. By employing a bias-adjusting to this $[X, X]_T^{(\text{avg})}$, the resulting consistent estimator is called the Two Time Scales estimator of the integrated volatility, and later renamed as Two Scales Realized Volatility (TSRV) in Aït-Sahalia et al. (2011). This latter reference, Aït-Sahalia et al. (2011), provides the results for dependent microstructure noise which is stationary and strong mixing with the mixing coefficients decaying exponentially, together with a moment condition for the microstructure. Also, they include explicit example of the AR(1) microstructure.

Extending these serial-dependent microstructure noise, in this section, we investigate the asymptotic distribution of realized volatility and its subsample estimator under the case that microstructure noise follows an ARCH model of order ∞ (ARCH(∞)) (Engle (1982); Giraitis et al. (2000)). In fact, empirical evidence shows that the process of market microstructure usually has the conditional heteroskedasticity (see Madhavan (2000), and we will discuss on this issue in Remark 4.4 below). The usual ARCH model itself may be included by the situation of Aït-Sahalia et al. (2011), since some conditions make it satisfy the strong mixing condition with geometric rate (see Francq and Zakoïan (2006); Meitz and Saikkonen (2008)). The distinction between our setting and that of Aït-Sahalia et al. (2011) will be described in Remark 4.3. Also, there is a merit when we specify the model for the microstructure, such as forecasting, etc. We also provide the numerical results, which show the MSE of lag averaging realized volatility is smaller than that of usual RV even if noise follows ARCH(∞).

4.3.1 Model, Estimators, and Main Results

In this section, first, we describe stochastic models for price, and explain the integrated and realized volatilities. Let $(\Omega, \mathcal{F}, P)$ be a probability space, and let $\{\mathcal{F}_t : 0 \le t \le \tau\}$ be a nondecreasing family of σ-algebra satisfying $\mathcal{F}_s \subseteq \mathcal{F}_t \subseteq \mathcal{F}, s \le t$. We suppose that the logarithmic price of a given asset $X = \{X_t, \mathcal{F}_t : 0 \le t \le \tau\}$ follows a $\{\mathcal{F}_t\}$-adapted continuous time diffusion process, that is, Ito's process:

$$dX_t = \mu_t dt + \sigma_t dW_t \tag{4.35}$$

where μ_t is the drift component, σ_t is the instantaneous volatility and $\{W_t, \mathcal{F}_t\}$ is the $\{\mathcal{F}_t\}$-adapted standard Brownian motion process.

Assumption 4.4 μ_t and σ_t are $\{\mathcal{F}_t\}$-adapted for each $0 \leq t \leq \tau$, and satisfy

$$P\left\{\int_0^\tau |\mu_t| dt < \infty\right\} = 1, \tag{4.36}$$

$$P\left\{\int_0^\tau \sigma_t^2 dt < \infty\right\} = 1. \tag{4.37}$$

In real trading, we can only use discretized data of assets though price process is assumed to follow continuous process (4.35). In what follows, we denote the t-th observation of i-th day by $X_{i+\frac{t}{M}}$, and write the true (latent) return as

$$r_{i+\frac{t}{M}} := X_{i+\frac{t}{M}} - X_{i+\frac{t-1}{M}},$$

where $1/M$ is the length of sampling interval. The Integrated Volatility was introduced to describe the true volatility of continuous latent process. The Integrated Volatility of the i-th day, $\langle X, X\rangle_{i,i+1}$, is defined as

$$\langle X, X\rangle_{i,i+1} := \int_i^{i+1} \sigma_t^2 dt. \tag{4.38}$$

A good estimator for $\langle X, X\rangle_{i,i+1}$ is the Realized Volatility of i-th day defined by

$$[X, X]_{i,i+1} := \sum_{t=1}^{M} r_{i+\frac{t}{M}}^2. \tag{4.39}$$

Then, the following results are given by Bandi and Russell (2008).

Proposition 4.3 (Bandi and Russell (2008)) *Under Assumption 4.4,*

$$\underset{M\to\infty}{p\text{-}\lim}([X, X]_{i,i+1} - \langle X, X\rangle_{i,i+1}) = 0, \tag{4.40}$$

$$\sqrt{M}([X, X]_{i,i+1} - \langle X, X\rangle_{i,i+1}) \xrightarrow{d} \mathcal{MN}(0, 2Q_{i,i+1}), \qquad M \to \infty, \tag{4.41}$$

where $\mathcal{MN}(0, 2Q_{i,i+1})$ *is the mixed normal distribution with mean 0 and random variance* $2Q_{i,i+1}$*, where*

$$Q_{i,i+1} := \int_i^{i+1} \sigma_t^4 dt. \tag{4.42}$$

Now, let us consider the case that $Y_{i+\frac{t}{M}}$ is a perturbed observation with microstructure noise corresponding to the true (latent) log price $X_{i+\frac{t}{M}}$. The relation can be modeled as

$$Y_{i+\frac{t}{M}} = X_{i+\frac{t}{M}} + \epsilon_{i+\frac{t}{M}}, \tag{4.43}$$

where $\{\epsilon_t\}$ is a sequence of microstructure noise. In this section we assume that the series $\{\epsilon_{i+\frac{t}{M}}, t \in \mathbf{Z}\}$ follows an autoregressive conditionally heteroskedastic process of order ∞ (ARCH(∞)), i.e.,

$$\epsilon_{i+\frac{t}{M}} = \left(a_0 + \sum_{j=1}^{\infty} a_j \epsilon^2_{i+\frac{t-j}{M}}\right)^{1/2} z_{i+\frac{t}{M}}, \tag{4.44}$$

where $a_0 > 0,\ a_j \geq 0\ (j \geq 1)$, and z_t's are i.i.d. random variables.

Remark 4.3 Zhang et al. (2005) dealt with the case when $\{\epsilon_t\} \sim$ i.i.d. Further, Aït-Sahalia et al. (2011) discussed the case when $\{\epsilon_t\}$ is a strong mixing process with the mixing coefficient decaying exponentially (i.e., j-lag autocovariance function is of order ρ^j, $|\rho| < 1$.) In this section we deal with ARCH(∞) in (4.44), which is uncorrelated but dependent. Here, we assume that the coefficients $\{a_j\}$ of our ARCH(∞) satisfy Assumption 4.5 in Theorem 4.2, and that $a_j \sim O(j^{-r}),\ r > 1$ in Theorem 4.3. Hence our model and that of Aït-Sahalia et al. (2011) show essentially different dependence structure. It may be noted that, the ARCH(∞) setting is convenient for description of heteroskedasticity in the volatility. □

Remark 4.4 This assumption of heteroskedastic microstructure is natural when considering that the market microstructure is caused by artificial operation such as bid-ask bounce and asynchronous trading. Empirical evidence also shows the process of market microstructure usually has the conditional heteroskedasticity (see Madhavan (2000)). For example, automated limit order books systems of the type used by the Toronto Stock Exchange and Paris Bourse offer continuous trading with high degrees of transparency (i.e., public display of current and away limit orders) without reliance on dealers. Also, this heteroskedastic noise may have a reasoning in view of the informational asymmetries (see e.g., O'Hara (1997)). That is, by this persistence of ARCH(∞) we are trying to describe how quickly the information is assimilated into the latent process of asset price. Here, the existence of new information is governed by the uncorrelated z_i's but once if they realized then its direction and impact of concerning information are believed to persist since they are not yet assimilated. There, the duration how long it takes to attain the full-information price (i.e., the latent price) is unspecified but described in terms of the significant order of ARCH coefficients. □

To guarantee the stationarity and geometrical ergodicity (see Giraitis et al. (2000)), we assume the following.

Assumption 4.5 The coefficients in (4.44) satisfy the conditions below:

$$a_0 > 0, \quad a_j \geq 0, \quad \sum_{j=1}^{\infty} a_j < 1, \tag{4.45}$$

$$E[z_0^2]^{1/2} \sum_{j=1}^{\infty} a_j < 1. \tag{4.46}$$

Perturbed version of the return is

$$\tilde{r}_{i+\frac{t}{M}} := Y_{i+\frac{t}{M}} - Y_{i+\frac{t-1}{M}} = r_{i+\frac{t}{M}} + \eta_{i+\frac{t}{M}}, \tag{4.47}$$

where

$$\eta_{i+\frac{t}{M}} := \epsilon_{i+\frac{t}{M}} - \epsilon_{i+\frac{t-1}{M}}.$$

The observed RV is

$$[Y, Y]_{i,i+1} = \sum_{t=1}^{M} \tilde{r}^2_{i+\frac{t}{M}}. \tag{4.48}$$

We can rewrite the realized volatility (4.48) as the sum of three components,

$$[Y, Y]_{i,i+1} = [X, X]_{i,i+1} + \sum_{t=1}^{M} \eta^2_{i+\frac{t}{M}} + 2\sum_{t=1}^{M} r_{i+\frac{t}{M}} \eta_{i+\frac{t}{M}} \tag{4.49}$$

$$= \mathcal{I}_1 + \mathcal{I}_2 + 2\mathcal{I}_3. \quad \text{(say)} \tag{4.50}$$

If the true price process was observable, only the term $\mathcal{I}_1$ would drive the limiting properties of $[Y, Y]_{i,i+1}$. The presence of microstructure noise introduces two additional components $\mathcal{I}_2$ and $\mathcal{I}_3$. We will show that it is mainly the term $\mathcal{I}_2$ that makes standard consistency arguments fail. Intuitively, $\mathcal{I}_2$ diverge to infinity almost surely as the number of observations increases (or equivalently, as the frequency of observations increases) since more and more noise will get accumulated. Under ARCH(∞) disturbances, we have the following theorem.

Theorem 4.2 *Suppose that Y follows the model (4.44) with X of (4.35), and that Assumptions 4.4 and 4.5 hold. Then, as $M \to \infty$, conditionally on the X process, we have*

$$\frac{1}{\sqrt{M}}([Y, Y]_{i,i+1} - [X, X]_{i,i+1} - 2Mv) \xrightarrow{d} N(0, \rho_1 + \rho_2),$$

where

$$\rho_1 := 4\int_{-\pi}^{\pi}\int_{-\pi}^{\pi}\int_{-\pi}^{\pi} f^{\epsilon}(\lambda_1, \lambda_2, -\lambda_2)d\lambda_1 d\lambda_2 d\lambda_3 + 12\upsilon^2,$$

$$\rho_2 := 4\int_{-\pi}^{\pi}\int_{-\pi}^{\pi}\int_{-\pi}^{\pi} \left\{ e^{i(\lambda_2+\lambda_3)} f^{\epsilon}(\lambda_1, \lambda_2, \lambda_3) - 2e^{-i\lambda_2} f^{\epsilon}(\lambda_1, \lambda_2, -\lambda_2) \right\} d\lambda_1 d\lambda_2 d\lambda_3.$$

Here,

$$\upsilon := E\left[\epsilon^2_{i+\frac{t}{M}}\right],$$

$$f^{\epsilon}(\lambda_1, \lambda_2, \lambda_3) := \frac{1}{(2\pi)^3} \sum_{l_1,l_2,l_3=-\infty}^{\infty} Q^{\epsilon}(l_1, l_2, l_3) e^{-i(\lambda_1 l_1 + \lambda_2 l_2 + \lambda_3 l_3)},$$

$$Q^{\epsilon}(l_1, l_2, l_3) := \mathrm{cum}\left(\epsilon_i, \epsilon_{i+\frac{l_1}{M}}, \epsilon_{i+\frac{l_2}{M}}, \epsilon_{i+\frac{l_3}{M}}\right).$$

Proof See Taniai et al. (2012, Theorem 1). □

This theorem implies that the observed RV, $[Y, Y]_{i,i+1}$, has a diverging bias $2M\upsilon$ around the true RV $[X, X]_{i,i+1}$ as in the case of i.i.d. noise. However, the asymptotics are affected by non-Gaussian dependent structure of the noise process. Indeed, when the ϵ is i.i.d.(see Bandi and Russell (2008)), our $f^{\epsilon}(\lambda_1, \lambda_2, \lambda_3)$ becomes a constant, hence, $\rho_2 = 0$. Thus, ρ_2 represents a deviation from i.i.d. assumption on $\{\epsilon_t\}$. Since the fourth-order cumulant spectral density $f^{\epsilon}(\lambda_1, \lambda_2, \lambda_3)$ shows the degree of dependence and non-Gaussianity of $\{\epsilon_t\}$, we can examine their influences on the asymptotics of $[Y, Y]_{i,i+1}$.

As for the estimator of υ above, Zhang et al. (2005) introduced $\tilde{\upsilon}$ defined by

$$\tilde{\upsilon} := \frac{\sum_{t=1}^{M} \tilde{r}^2_{i+\frac{t}{M}}}{2M} = \frac{[Y, Y]_{i,i+1}}{2M}.$$

Accordingly, they proposed a bias-adjusted estimator for $\langle X, X\rangle_{i,i+1}$ by

$$[Y, Y]^{(\mathrm{tsrv})}_{i,i+1} := [Y, Y]^{(\mathrm{avg})}_{i,i+1} - \frac{\bar{M}}{M}[Y, Y]_{i,i+1}, \qquad \text{(: Two Scales Realized Volatility (tsrv))} \tag{4.51}$$

where $[Y, Y]^{(\mathrm{avg})}_{i,i+1}$ is defined, in the manner of (4.34), as

$$[Y, Y]^{(\mathrm{avg})}_{i,i+1} := \frac{1}{K}\sum_{k=1}^{K}[Y, Y]^{(k)}_{i,i+1},$$

$$[Y, Y]^{(k)}_{i,i+1} := \sum_{t=1}^{n_k}\left(Y_{i+\frac{k-1+tK}{M}} - Y_{i+\frac{k-1+(t-1)K}{M}}\right)^2 \quad \text{with} \quad n_k := \left\lfloor \frac{M-k+1}{K} \right\rfloor, \tag{4.52}$$

and $\bar{M}$ is

$$\bar{M} := \frac{1}{K}\sum_{k=1}^{K} n_k \approx \frac{M+1}{K} - \frac{K+1}{2} \approx \frac{M}{K},$$

where the last approximation is used when M is much larger relatively to K. The following theorem of ours states the result for ARCH(∞) disturbances case.

Theorem 4.3 *Suppose that Y follows the model (4.43) with X of (4.35), and that Assumptions 4.4 and 4.5 hold. Further, if the coefficients a_j's satisfy $a_j < bj^{-r}$ for some $r > 1$ and $b > 0$, and if $K = cM^{2/3}$, for some $c > 0$, then it holds that*

$$M^{1/6}([Y, Y]^{(\mathrm{tsrv})}_{i,i+1} - \langle X, X\rangle_{i,i+1}) \xrightarrow{d} \mathcal{MN}(0, c^{-2}(4\gamma + 12\upsilon^2)) + \mathcal{MN}(0, c(\eta^2))$$
$$= \mathcal{MN}(0, c^{-2}(4\gamma + 12\upsilon^2) + c(\eta^2)),$$

where

$$\eta := \frac{4}{3}\int_i^{i+1} \sigma_t^4 dt,$$
$$\gamma := \int_{-\pi}^{\pi}\int_{-\pi}^{\pi}\int_{-\pi}^{\pi} e^{i(\lambda_2+\lambda_3)} f^{\epsilon}(\lambda_1, \lambda_2, \lambda_3) d\lambda_1 d\lambda_2 d\lambda_3.$$

Proof See Taniai et al. (2012, Theorem 2). □

Theorem 4.3 implies the modified estimator $[Y, Y]^{(\mathrm{tsrv})}_{i,i+1}$ is consistent for $\langle X, X\rangle_{i,i+1}$ even if the noise process is ARCH(∞). This is because the subsampling frequency K also diverges to infinity when the original sampling frequency M goes to infinity. It may be noted that the asymptotics of $[Y, Y]^{(\mathrm{tsrv})}_{i,i+1}$ depend on non-Gaussianity and dependence of the noise through the integral of fourth-order cumulant spectra, which describes the asymptotics of the Whittle estimator for non-Gaussian dependent processes in unified manner (see, Hosoya and Taniguchi (1982), Sect. 3.1 of Taniguchi and Kakizawa (2000)).

Remark 4.5 Estimation for the asymptotic variances γ and υ^2 seems important. It is difficult to develop the general theory. But, a partial solution is given as follows. Suppose that $\{X_t\}$ is generated by a stationary continuous time ARMA process (See Brockwell (1994), and also Taniguchi and Kakizawa (2000, p. 149)), which includes a simple diffusion process whose discretized version is an autoregressive model. Consider

$$Y_{t_j} = X_{t_j} + \epsilon_{t_j}, \quad j = 1, 2, \ldots, n. \tag{4.53}$$

Based on the observation $\{Y_{t_j}\}$ we can estimate the spectral density $f_Y(\lambda)$ of $\{Y_{t_j}\}$. By using a method due to Hosoya and Taniguchi (1982, pp. 139–140), we can estimate the spectral density $f_X(\lambda)$ of $\{X_{t_j}\}$, and derive an estimator $\hat{X}_{t_j}$ of X_{t_j}. Then we can calculate $\hat{\epsilon}_{t_j} \equiv Y_{t_j} - \hat{X}_{t_j}$. Based on $\hat{\epsilon}_{t_1}, \ldots, \hat{\epsilon}_{t_n}$, we can estimate the integral

$$\gamma = \int_{-\pi}^{\pi}\int_{-\pi}^{\pi}\int_{-\pi}^{\pi} e^{i(\lambda_2+\lambda_3)} f^{\epsilon}(\lambda_1, \lambda_2, \lambda_3) d\lambda_1 d\lambda_2 d\lambda_3.$$

by a method proposed by Taniguchi (1982). Hence we can construct a consistent estimator $\hat{\gamma}$ of γ.

4.3.2 Numerical Studies

In this section, we calculate two estimators $[Y, Y]_{i,i+1}$, $[Y, Y]_{i,i+1}^{(\mathrm{tsrv})}$ to estimate the integrated volatility $\langle X, X\rangle_{i,i+1}$. The observed frequency M is chosen as 25,000, which corresponds to the time interval about 1 s. In order to generate the X process, here we use the Euler-Maruyama Discretization (e.g., Kloeden and Platen (1992); Gouriéroux and Jasiak (2001)):

$$X_{i+\frac{t}{M}} = X_{i+\frac{t-1}{M}} + \mu_{i+\frac{t-1}{M}}\frac{1}{M} + \sigma_{i+\frac{t-1}{M}}\left(W_{i+\frac{t}{M}} - W_{i+\frac{t-1}{M}}\right).$$

So we may express the log-return process as

$$r_{i+\frac{t}{M}} = X_{i+\frac{t}{M}} - X_{i+\frac{t-1}{M}} = \mu_{i+\frac{t-1}{M}}\frac{1}{M} + \sigma_{i+\frac{t-1}{M}}\sqrt{\frac{1}{M}}u_{i+\frac{t}{M}}, \tag{4.54}$$

with $\{u_t\}$ being a sequence of i.i.d. $N(0, 1)$ random variables. Consequently, provided the condition (4.37), a discretized version of Integrated Volatility (4.38), which here we define as

$$\langle X, X\rangle_{i,i+1}^{(d)} := \frac{1}{M}\sum_{t=1}^{M}\sigma^2_{i+\frac{t-1}{M}}, \tag{4.55}$$

is known to converge to $\langle X, X\rangle_{i,i+1}$ as $M \to \infty$. Now, let us assume that $\mu_{i+\frac{t}{M}}$ and $\sigma_{i+\frac{t}{M}}$ take the form as

$$\mu_{i+\frac{t}{M}} = \phi_0 + \phi_1 r_{i+\frac{t-1}{M}}, \qquad \sigma^2_{i+\frac{t}{M}} = s_0 + s_1(r_{i+\frac{t-1}{M}} - \mu_{i+\frac{t-1}{M}})^2,$$

so that the log-return $r_{i+\frac{t}{M}}$ follows an AR(1)-ARCH(1) model characterized by

$$\begin{aligned} r_{i+\frac{t}{M}} &= \frac{\phi_0}{M} + \frac{\phi_1}{M} r_{i+\frac{t-1}{M}} + e_{i+\frac{t}{M}}, \quad e_{i+\frac{t}{M}} := h_{i+\frac{t}{M}} u_{i+\frac{t}{M}}, \\ h^2_{i+\frac{t}{M}} &= \frac{s_0}{M} + \frac{s_1}{M} e^2_{i+\frac{t-1}{M}}. \end{aligned} \tag{4.56}$$

Then, we generate the process $\{X_{i+\frac{t}{M}}\}$ through the above AR-ARCH model (4.56) with a choice of parameters as follows:

$$\begin{aligned} r_{i+\frac{t}{M}} &= 10^{-5} + 0.3 r_{i+\frac{t-1}{M}} + e_{i+\frac{t}{M}}, \quad e_{i+\frac{t}{M}} := h_{i+\frac{t}{M}} u_{i+\frac{t}{M}}, \\ h^2_{i+\frac{t}{M}} &= \frac{s_0}{M} + 0.5 e^2_{i+\frac{t-1}{M}}. \end{aligned} \tag{4.57}$$

Here, according to our choice $s_1/M \equiv 0.5$, the value of s_0 is set to be $s_0 \equiv 1 - s_1/M$ so that the unconditional standard deviation over 1 period $(1/M)$ becomes $(\mathrm{Var}[r_{i+\frac{t}{M}}])^{1/2} \equiv 1/\sqrt{M}$. Also, we assume the market microstructure noise $\{\epsilon_t\}$ follows an ARCH(1) process

$$\epsilon_{i+\frac{t}{M}} = \left(a_0 + a_1 \epsilon^2_{i+\frac{t-1}{M}}\right)^{1/2} \sqrt{\frac{1}{M}} z_{i+\frac{t}{M}} = \left(\frac{a_0}{M} + \frac{a_1}{M} \epsilon^2_{i+\frac{t-1}{M}}\right)^{1/2} z_{i+\frac{t}{M}}. \tag{4.58}$$

In our first observation below, we set $a_0 \equiv p^2 - a_1/M$ for varying values of a_1 to keep the unconditional standard deviation to be $(\mathrm{Var}[\epsilon_{i+\frac{t}{M}}])^{1/2} \equiv p/\sqrt{M}$. That is, the microstructure noise is set to always have $100p\%$ magnitude with respect to the asset. Substituting $u_{i+\frac{t}{M}}$ in (4.54) and $z_{i+\frac{t}{M}}$ in (4.58) with realizations of $N(0, 1)$, we generate $20 \times M$ (which amounts to 1-month data points) samples of $\{X_{i+\frac{t}{M}}\}$.

Let $\mathrm{MSE}^{(\mathrm{rv})}$ and $\mathrm{MSE}^{(\mathrm{tsrv})}$ denote the mean squared error (MSE) of $[Y, Y]_{i,i+1}$ and $[Y, Y]^{(\mathrm{tsrv})}_{i,i+1}$, i.e.,

$$\begin{aligned} \mathrm{MSE}^{(\mathrm{rv})} &:= \frac{1}{20} \sum_{i=1}^{20} ([Y, Y]_{i,i+1} - \langle X, X\rangle^{(d)}_{i,i+1})^2, \\ \mathrm{MSE}^{(\mathrm{tsrv})} &:= \frac{1}{20} \sum_{i=1}^{20} ([Y, Y]^{(\mathrm{tsrv})}_{i,i+1} - \langle X, X\rangle^{(d)}_{i,i+1})^2. \end{aligned}$$

Figure 4.5 shows the values of $\mathrm{MSE}^{(\mathrm{tsrv})}$ with $p = 1$ for several choices of sampling interval K and ARCH parameter a_1.

As we will see in Table 4.3, for this set of parameters the value of $\mathrm{MSE}^{(\mathrm{rv})}$ is about 4.42763. So, it can be seen that the TSRV estimator yields smaller estimation error with any choice of the sampling interval from $K = 2$ (2 s) to $K = 600$ (10 min.). Also, it should be noted that $\mathrm{MSE}^{(\mathrm{tsrv})}$ is almost robust against the choice of the ARCH parameter. This feature of TSRV is constrast to those of RV, so we study the following Table 4.3 with this in our mind.

Table 4.3 shows the values of $\mathrm{MSE}^{(\mathrm{rv})}$ and $\mathrm{MSE}^{(\mathrm{tsrv})}$ for $K = 10$, along the choices of the noise magnitude p and ARCH parameter a_1. As mentioned, there we see that $\mathrm{MSE}^{(\mathrm{rv})}$ is more sensitive to the change of ARCH parameter than $\mathrm{MSE}^{(\mathrm{tsrv})}$. From this Table 4.3, we may find two features. Firstly, the defference between $\mathrm{MSE}^{(\mathrm{rv})}$ and $\mathrm{MSE}^{(\mathrm{tsrv})}$ varies along the choice of the size of noise p. In fact, if the relative size of $\epsilon_{i+\frac{t}{M}}$ is small enough with respect to the asset $X_{i+\frac{t}{M}}$, we even observe $\mathrm{MSE}^{(\mathrm{rv})}$ outperforming $\mathrm{MSE}^{(\mathrm{tsrv})}$ uniformly in a_1 (and in K also). This may be related to our second remark of Table 4.3, which is the fact that those MSEs exhibit a decreasing

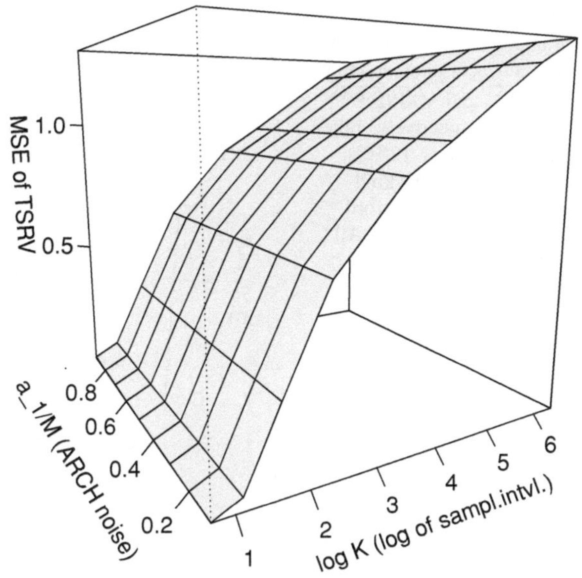

Fig. 4.5 $\text{MSE}^{(\text{tsrv})}$ with $p = 1$. Sampling interval $\log K$ are log(2, 3, 5, 10, 30, 60, 300, 600). As a comparison, $\text{MSE}^{(\text{rv})}$ is larger than 2.5 for $a_1/M \in [0.1, 0.9]$, as shown in Table 4.3. Taken from Taniai et al. (2012). Published with the kind permission of © Oxford University Press 2012. All Rights Reserved

Table 4.3 The values of $\text{MSE}^{(\text{rv})}$ (upper) and $\text{MSE}^{(\text{tsrv})}$ (lower) when $K = 10$

		ARCH parameter a_1/M				
		0.1	0.3	0.5	0.7	0.9
Noise magnitude	0	0.00987	0.00987	0.00987	0.00987	0.00987
p		0.61247	0.61247	0.61247	0.61247	0.61247
	0.2	0.03210	0.03208	0.03199	0.03153	0.02579
		0.61271	0.61272	0.61280	0.61295	0.61298
	0.4	0.17595	0.17577	0.17496	0.17085	0.12040
		0.61302	0.61305	0.61318	0.61348	0.61346
	0.6	0.67318	0.67237	0.66891	0.65122	0.43442
		0.61340	0.61344	0.61363	0.61405	0.61394
	0.8	1.90997	1.90755	1.89733	1.84492	1.20249
		0.61384	0.61391	0.61414	0.61468	0.61440
	1	4.42692	4.42117	4.39710	4.27319	2.75318
		0.61435	0.61444	0.61471	0.61535	0.61486

Taken from Taniai et al. (2012). Published with the kind permission of © Oxford University Press 2012. All Rights Reserved

feature in a_1. Suspecting that this stems from our request $a_0 \equiv p^2 - a_1/M$, which was to keep the unconditional variance, we investigate the behavior of MSEs along the choice of ARCH parameters both a_0 and a_1 below.

By releasing the constraint $a_0 = p^2 - a_1/M$, Fig. 4.6 exhibits the changes of MSEs in ARCH parameters a_0 and a_1. Again, as for the sampling interval of TSRV estimator, we applied $K = 10$. We observe that $\text{MSE}^{(\text{rv})}$, the upper surface, is more sensitive than $\text{MSE}^{(\text{tsrv})}$, the lower surface, to both parameters a_0 and a_1. In fact, the

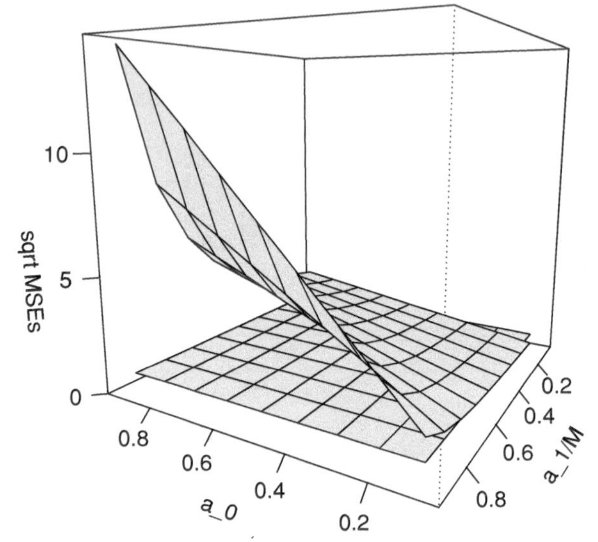

Fig. 4.6 $\sqrt{\text{MSE}}$s for ARCH parameters. The *upper* and *lower* surfaces coorespond to $\sqrt{\text{MSE}^{(\text{rv})}}$ and $\sqrt{\text{MSE}^{(\text{tsrv})}}$ ($K = 10$) respectively. Taken from Taniai et al. (2012). Published with the kind permission of © Oxford University Press 2012. All Rights Reserved

lower surface $\text{MSE}^{(\text{tsrv})}$ is almost a horizontal plane, which shows the robustness of TSRV estimator against the ARCH affection. This, in turn, suggest the importance of recognition of ARCH behavior in microstructure noise when applying the commonly used RV estimator (4.32).

4.4 Appendix

4.4.1 Proof of Theorem 4.1

To begin with, define, in regard to (4.13),

$$\boldsymbol{D}^{(n)}_{\boldsymbol{\gamma},f} := \frac{1}{n-p} \sum_{t=p+1}^{n} E^{(n)}_{\boldsymbol{\gamma},f}[\boldsymbol{\psi}_{f,p}(U_{\boldsymbol{\gamma},t}, \ldots, U_{\boldsymbol{\gamma},t-p}) \,|\mathcal{B}^{(n)}_{\boldsymbol{\gamma}}].$$

Then, as in Hallin et al. (2008, Equation (A.4)), it can be shown that

$$\boldsymbol{D}^{(n)}_{\boldsymbol{\gamma},f} - E^{(n)}_{\boldsymbol{\gamma},f}[\boldsymbol{D}^{(n)}_{\boldsymbol{\gamma},f} \,|N^{(n)}_{\boldsymbol{\gamma},L}] = \frac{1}{n-p} \sum_{t=p+1}^{n} \Big[\boldsymbol{\psi}_{f,p}(U_{\boldsymbol{\gamma},t}, \ldots, U_{\boldsymbol{\gamma},t-p}) - E[\boldsymbol{\psi}_{f,p}(U_0, \ldots, U_p) \,|U_0 = U_{\boldsymbol{\gamma},t}]\Big] + o_P(n^{-1/2}). \tag{4.59}$$

More precisely, first, with help of results for Linear Serial Rank Statistics (See Hallin et al. (2003, Sect. 4.2)), it can be shown that

$$\boldsymbol{D}^{(n)}_{\gamma,f} - E^{(n)}_{\gamma,f}[\boldsymbol{D}^{(n)}_{\gamma,f}\,|N^{(n)}_{\gamma,L}] = \boldsymbol{T}^{(n)}_{\gamma,f} - E[\boldsymbol{T}^{(n)}_{\gamma,f}\,|\boldsymbol{\xi}_{\gamma,(\cdot)}] + o_P(n^{-1/2}), \tag{4.60}$$

where

$$\boldsymbol{T}^{(n)}_{\gamma,f} := \frac{1}{n-p}\sum_{t=p+1}^{n} \boldsymbol{\psi}_{f,p}(U_{\gamma,t},\ldots,U_{\gamma,t-p}).$$

and $\boldsymbol{\xi}_{\gamma,(\cdot)}$ denotes the vector of Order Statistics. Then, by van der Vaart (1998, Theorem 12.3) or Serfling (1980, Sect. 5.3), the second term of the r.h.s. of (4.60) becomes into

$$\begin{aligned} E[\boldsymbol{T}^{(n)}_{\gamma,f}|\boldsymbol{\xi}_{\gamma,(\cdot)}] &= E[\boldsymbol{T}^{(n)}_{\gamma,f}|\boldsymbol{U}_{\gamma,(\cdot)}] \\ &= \frac{1}{n-p}\sum_{t=p+1}^{n}\sum_{\ell=0}^{p} E[\boldsymbol{\psi}_{f,p}(U_0,\ldots,U_p)|U_l = U_{\gamma,t-\ell}] \\ &\quad + (p+1)E[\boldsymbol{\psi}_{f,p}(U_0,\ldots,U_p)] + o_P(n^{-1/2}). \end{aligned}$$

So, together with the fact that $\boldsymbol{\psi}_{f,p}$ forms a martingale difference, that is, for all $(u_1,\ldots,u_p) \in [0,1]^p$,

$$\int_0^1 \boldsymbol{\psi}_{f,p}(u_0,u_1,\ldots,u_p)\mathrm{d}u_0 = \boldsymbol{0}, \tag{4.61}$$

we have (4.59).

Evaluation of $E^{(n)}_{\gamma,f}[\boldsymbol{D}^{(n)}_{\gamma,f}\,|N^{(n)}_{\gamma,L}]$ is also a modification of Hallin et al. (2008). That is, firstly observe that, by the tower property of conditional expectations and WLLN,

$$\begin{aligned} E^{(n)}_{\gamma,f}[\boldsymbol{D}^{(n)}_{\gamma,f}\,|N^{(n)}_{\gamma,L}] &= \frac{1}{n-p}\sum_{t=p+1}^{n} E^{(n)}_{\gamma,f}[\boldsymbol{\psi}_{f,p}(U_{\gamma,t},\ldots,U_{\gamma,t-p})\,|N^{(n)}_{\gamma,L}] \\ &= E[\boldsymbol{\psi}^{(s)}_{f,p}(\boldsymbol{s})\,|N^{(n)}_{\gamma,L}] + o_P(n^{-1/2}), \end{aligned}$$

where $\boldsymbol{s} = (s_0,\ldots,s_p) \in \{-1,1\}^{p+1}$ and

$$\boldsymbol{\psi}^{(s)}_{f,p}(\boldsymbol{s}) := E^{(n)}_{\gamma,f}[\boldsymbol{\psi}_{f,p}(U_{\gamma,t},\ldots,U_{\gamma,t-p})\,|S_{\gamma,t} = s_0,\ldots,S_{\gamma,t-p} = s_p].$$

The number of -1's in its argument is distributed as

$$\#\{i = 0, 1, \dots, p : S_{\gamma,t-i} = -1\} \overset{d}{\sim} \text{HyperGeometric}(n,\ N^{(n)}_{\gamma,L},\ p),$$

so we approximate this with binomial distribution. Then, noting that $E^{(n)}_{\gamma,f}[\boldsymbol{D}^{(n)}_{\gamma,f}] = 0$ by (4.61), we have

$$\begin{aligned}
&E^{(n)}_{\gamma,f}[\boldsymbol{D}^{(n)}_{\gamma,f}\ |N^{(n)}_{\gamma,L}] - E^{(n)}_{\gamma,f}[\boldsymbol{D}^{(n)}_{\gamma,f}] \\
&= \sum_{s\in\{-1,1\}^{p+1}} \boldsymbol{\psi}^{(s)}_{f,p}(\boldsymbol{s}) \cdot B_p\left(\frac{N^{(n)}_{\gamma,L}}{n}, \#\{i : s_i = -1\}\right) \\
&\quad - \sum_{s\in\{-1,1\}^{p+1}} \boldsymbol{\psi}^{(s)}_{f,p}(\boldsymbol{s}) \cdot B_p(\tau, \#\{i : s_i = -1\}) + o_P(n^{-1/2}) \\
&= \sum \boldsymbol{\psi}^{(s)}_{f,p}(\boldsymbol{s}) \cdot \frac{\partial}{\partial x} B_p(x, \#\{i : s_i = -1\})\Big|_{x=\tau} \left(\frac{N^{(n)}_{\gamma,L}}{n} - \tau\right) + o_P(n^{-1/2}) \\
&= \sum \boldsymbol{\psi}^{(s)}_{f,p}(\boldsymbol{s}) \cdot B_p(\tau, \#\{s_i = -1\}) \frac{\#\{i : s_i = -1\} - \tau(p+1)}{\tau(1-\tau)} \\
&\quad \times \left(\frac{N^{(n)}_{\gamma,L}}{n} - \tau\right) + o_P(n^{-1/2}) \\
&= \frac{E[\boldsymbol{\psi}^{(s)}_{f,p}(\boldsymbol{s})(\#\{i : s_i = -1\} - \tau(p+1))]}{\tau(1-\tau)} \times \left(\frac{N^{(n)}_{\gamma,L}}{n} - \tau\right) + o_P(n^{-1/2}),
\end{aligned} \tag{4.62}$$

where $B_p(x, w) := x^w(1-x)^{p+1-w}$. Now, observe that

$$\begin{aligned}
&E[\boldsymbol{\psi}^{(s)}_{f,p}(\boldsymbol{s})(\#\{i = 0, 1, \dots, p : s_i = -1\} - \tau(p+1))\ |s_1, \dots, s_p] \\
&= P\{s_0 = -1\}\boldsymbol{\psi}^{(s)}_{f,p}(-1, s_1, \dots, s_p)(1 + \#\{i = 1, \dots, p : s_i = -1\} - \tau(p+1)) \\
&\quad + P\{s_0 = 1\}\boldsymbol{\psi}^{(s)}_{f,p}(1, s_1, \dots, s_p)(0 + \#\{i = 1, \dots, p : s_i = -1\} - \tau(p+1)) \\
&= \tau \cdot \boldsymbol{\psi}^{(s)}_{f,p}(-1, s_1, \dots, s_p),
\end{aligned}$$

where the last equality follows from $P\{s_0 = -1\} = P\{S_{\gamma,t} = -1\} = \tau$ and

$$P\{s_0 = -1\}\boldsymbol{\psi}^{(s)}_{f,p}(-1, s_1, \dots, s_p) + P\{s_0 = 1\}\boldsymbol{\psi}^{(s)}_{f,p}(1, s_1, \dots, s_p) = 0,$$

by (4.61). The tower property yields that

$$
\begin{aligned}
&E[\psi^{(s)}_{f,p}(s)\big(\#\{i : s_i = -1\} - \tau(p+1)\big)] \\
&= \tau \cdot E[\psi^{(s)}_{f,p}(s) \,|s_0 = -1] = \tau \cdot E[\psi_{f,p}(U_0, \ldots, U_p) \,|U_0 \le \tau] \\
&= \tau \cdot \frac{\int_0^\tau \varphi_f[F^{-1}(u)]du}{\int_0^\tau du} \cdot \mu_{W(\gamma,f)} = -f(1) \cdot \mu_{W(\gamma,f)}.
\end{aligned}
$$

Finally, the rest of (4.62) is $N^{(n)}_{\gamma,L}/n - \tau = n^{-1} \sum_{t=1}^{n} (1\{\xi_{\gamma,t} \le 1\} - \tau)$.

The asymptotic normality (4.16) follows from Yoshihara's CLT for U-statistics, as shown in Hallin et al. (1985). The proof is now completed. □

References

Aït-Sahalia, Y., Mykland, P.A., Zhang, L.: Ultra high frequency volatility estimation with dependent microstructure noise. J. Econometric **160**(1), 160–175 (2011)

Andersen, T.G., Bollerslev, T., Diebold, F.X., Labys, P.: Modeling and forecasting realized volatility. Econometrica **71**(2), 579–625 (2003)

Bandi, F.M., Russell, J.R.: Microstructure noise, realized variance, and optimal sampling. Rev. Econom. Stud. **75**(2), 339–369 (2008)

Barndorff-Nielsen, O.E., Shephard, N.: Econometric analysis of realized volatility and its use in estimating stochastic volatility models. J. R. Stat. Soc. Ser. B Stat. Methodol. **64**(2), 253–280 (2002)

Bickel, P.J., Klaassen, C.A.J., Ritov, Y.: Efficient and Adaptive Estimation for Semiparametric Models. Springer, New York (1998). Reprint of the 1993 original

Brockwell, P.J.: On continuous-time threshold ARMA processes. J. Statist. Plann. Infer. **39**(2), 291–303 (1994)

Dacorogna, M., Gençay, R., Muller, U.A., Olsen, R., Pictet, O.: An Introduction to High-Frequency Finance. Academic, San Diego (2001)

Drost, F.C., Klaassen, C.A.J., Werker, B.J.M.: Adaptive estimation in time-series models. Ann. Statist. **25**(2), 786–817 (1997)

Engle, R.F.: Autoregressive conditional heteroscedasticity with estimates of the variance of United Kingdom inflation. Econometrica **50**(4), 987–1007 (1982)

Francq, C., Zakoïan, J.M.: Mixing properties of a general class of GARCH(1,1) models without moment assumptions on the observed process. Econometric Theor. **22**(5), 815–834 (2006)

Giraitis, L., Kokoszka, P., Leipus, R.: Stationary ARCH models: dependence structure and central limit theorem. Econometric Theor. **16**(1), 3–22 (2000)

Gouriéroux, C.: ARCH Models and Financial Applications. Springer, New York (1997). Springer Series in Statistics

Gouriéroux, C., Jasiak, J.: Financial Econometrics. Princeton University Press, Princeton (2001). Princeton-Series in Finance

Hallin, M., Ingenbleek, J.F., Puri, M.L.: Linear serial rank tests for randomness against ARMA alternatives. Ann. Statist. **13**(3), 1156–1181 (1985)

Hallin, M., Oja, H., Paindaveine, D.: Semiparametrically efficient rank-based inference for shape. II. Optimal R-estimation of shape. Ann. Statist. **34**(6), 2757–2789 (2006a)

Hallin, M., Vermandele, C., Werker, B.J.M.: Serial and nonserial sign-and-rank statistics: asymptotic representation and asymptotic normality. Ann. Statist. **34**(1), 254–289 (2006b)

Hallin, M., Vermandele, C., Werker, B.J.M.: Semiparametrically efficient inference based on signs and ranks for median-restricted models. J. R. Stat. Soc. Ser. B Stat. Methodol. **70**(2), 389–412 (2008)

Hallin, M., Werker, B.J.M.: Semi-parametric efficiency, distribution-freeness and invariance. Bernoulli **9**(1), 137–165 (2003)

He, C., Teräsvirta, T.: Fourth moment structure of the GARCH(p, q) process. Econometric Theory **15**(6), 824–846 (1999)

Hosoya, Y., Taniguchi, M.: A central limit theorem for stationary processes and the parameter estimation of linear processes. Ann. Statist. **10**(1), 132–153 (1982)

Kloeden, P.E., Platen, E.: Numerical solution of stochastic differential equations, Applications of Mathematics (New York), vol. 23. Springer, Berlin (1992)

Koenker, R.: Quantile regression, Econometric Society Monographs, vol. 38. Cambridge University Press, Cambridge (2005)

Koenker, R., Bassett Jr, G.: Regression quantiles. Econometrica **46**(1), 33–50 (1978)

Koenker, R., Xiao, Z.: Quantile autoregression. J. Amer. Statist. Assoc. **101**(475), 980–990 (2006)

Koul, H.L.: Weighted Empiricals and Linear Models. Institute of Mathematical Statistics Lecture Notes-Monograph Series, 21. Institute of Mathematical Statistics, Hayward, CA (1992).

Kreiss, J.P.: On adaptive estimation in stationary ARMA processes. Ann. Statist. **15**(1), 112–133 (1987)

Linton, O.: Adaptive estimation in ARCH models. Econometric Theor. **9**(4), 539–569 (1993)

Madhavan, A.: Market microstructure: A survey. J. Finan. Markets. **3**(3), 205–258 (2000)

McAleer, M., Medeiros, M.: Realized volatility: a review. Econometric Rev. **27**(1—-3), 10–45 (2008)

Meitz, M., Saikkonen, P.: Ergodicity, mixing, and existence of moments of a class of Markov models with applications to GARCH and ACD models. Econometric Theor. **24**(5), 1291–1320 (2008)

O'Hara, M.: Market Microstructure Theory. Blackwell, Cambridge, MA (1997)

Puri, M.L., Sen, P.K.: Nonparametric Methods in Multivariate Analysis. Wiley, New York-London-Sydney (1971)

Schmetterer, L.: Introduction to Mathematical Statistics. Springer, Berlin (1974). Translated from the second German edition by Kenneth Wickwire, Die Grundlehren der mathematischen Wissenschaften, Band 202

Serfling, R.J.: Approximation Theorems of Mathematical Statistics. Wiley, New York (1980). Wiley Series in Probability and Mathematical Statistics

Taniai, H.: Inference for the quantiles of ARCH processes. The doctoral dissertation. Université libre de Bruxelles, Belgium (2009)

Taniai, H., Usami, T., Suto, N., Taniguchi, M.: Asymptotics of realized volatility with non-gaussian arch(∞) microstructure noise. J. Finan. Econometrics **10**(4), 617–636 (2012)

Taniguchi, M.: On estimation of the integrals of the fourth order cumulant spectral density. Biometrika **69**(1), 117–122 (1982)

Taniguchi, M., Kakizawa, Y.: Asymptotic Theory of Statistical Inference for Time Series. Springer, New York (2000). Springer Series in Statistics

Tsay, R.S.: Analysis of Financial Time Series. Wiley, New York (2002). Wiley series in probability and mathematical statistics

van der Vaart, A.W.: Asymptotic Statistics. Cambridge University Press, Cambridge (1998). Cambridge Series in Statistical and Probabilistic Mathematics

Zhang, L., Mykland, P.A., Aït-Sahalia, Y.: A tale of two time scales: determining integrated volatility with noisy high-frequency data. J. Amer. Statist. Assoc. **100**(472), 1394–1411 (2005)

Index

A
American option, 29
Arbitrage opportunity, 28
Arbitrage-free, 28
ARCH model, 85, 105
ARCH(∞) model, 19
ARCH(∞)-SM model, 19
ARCH(q), 3
Asian call option, 30
Asset, 27
Asymptotic relative efficiency (ARE), 96
Asymptotically centering, 23
Asymptotically efficient, 23, 27

B
Black-Scholes formula, 32

C
Call option, 29
CAPM, 74
Central sequence, 22
Characteristic exponent, 53
CHARN, 5
Classification, 35
Complete, 29
Conditional LSE, 24
Conditional quantile, 86
Contiguous alternative, 25
Contingent claim, 29
Control variate estimator, 66
Control variate method, 66
CR statistic, 46
  frequency domain—, 47
Curved structure, 78

D
Dendrogram, 37
Discretization, 93
Disparity measure, 18

E
Edgeworth expansion, 32
EGARCH(p, q), 4
Empirical likelihood ratio function, 42
  —for frequency domain, 44
Equivalent martingale measure, 29
Estimating function, 42
European option, 29

F
Fourth-order cumulant, 6
Fourth-order stationary, 6
Fractionally cointegrated, 78

G
GARCH(p, q), 4
Generalized empirical likelihood (GEL), 46, 49
Geometric Brownian motion, 31
Grenander's condition, 71

H
h-step ahead prediction, 12

I
Innovation-free, 12
Instrumental variable, 78

SpringerBriefs in Statistics, DOI: 10.1007/978-3-319-03497-3, © The Author(s) 2014

Instrumental variable method, 75
Integrated volatility (IV), 102

K
Kolmogorov's formula, 46

L
LAN, 19
Locally stationary, 26
Log-likelihood ratio, 25
Long memory dependence, 74
Long memory dependent, 77

M
Market microstructure, 102
Martingale, 28
Maturity, 29
Maximal invariant, 89
Maximum likelihood estimator, 25
Mean squared error (MSE), 101
Measure of instantaneous linear feedback, 16
Measure of linear dependence, 17
Measure of linear feedback, 16
Misclassification, 35

N
Non-Gaussian robustness, 36
Nonparametric likelihood, 42
Nonparametric spectral estimator, 14

O
One-step estimator, 92
Option, 29
Ordinary least squares (OLS) estimator, 75, 78

P
Portfolio, 13, 27
Put option, 29

Q
Quantile regression (QR), 85
Quasi-Gaussian maximum likelihood estimator, 9
Quasi-maximum likelihood estimator, 27

R
Realized volatility (RV), 85, 102

S
Sample autocorrelation function (SACF), 2
Sample mean, 66, 70
Second-order stationary, 5
Self-financing portfolio, 29
Semiparametric efficiency, 95
Short memory dependent, 77
Short memory process, 75
Spectral measure, 53
Spectral moment condition, 43
Spectral window, 13
Stable distribution, 51
  multivariate—, 51
Stochastic regression model, 74, 76
Strike price, 29
SV(m), 4

T
The least squares estimator (LSE), 67, 72
Time varying spectral density, 26
Trend model, 70
Two scales realized volatility (TSRV), 103
Two-stage least squares (2SLS) estimator, 75, 79

U
ULAN, 90

W
Wold decomposition theorem, 5